Lecture Notes in Computer Science 15569

Founding Editors

Gerhard Goos
Juris Hartmanis

Editorial Board Members

Elisa Bertino, *Purdue University, West Lafayette, IN, USA*
Wen Gao, *Peking University, Beijing, China*
Bernhard Steffen, *TU Dortmund University, Dortmund, Germany*
Moti Yung, *Columbia University, New York, NY, USA*

The series Lecture Notes in Computer Science (LNCS), including its subseries Lecture Notes in Artificial Intelligence (LNAI) and Lecture Notes in Bioinformatics (LNBI), has established itself as a medium for the publication of new developments in computer science and information technology research, teaching, and education.

LNCS enjoys close cooperation with the computer science R & D community, the series counts many renowned academics among its volume editors and paper authors, and collaborates with prestigious societies. Its mission is to serve this international community by providing an invaluable service, mainly focused on the publication of conference and workshop proceedings and postproceedings. LNCS commenced publication in 1973.

Jordane Lorandel · Ahmed Kamaleldin
Editors

Design and Architecture for Signal and Image Processing

18th International Workshop, DASIP 2025
Barcelona, Spain, January 20–22, 2025
Proceedings

Editors
Jordane Lorandel
University of Rennes
Rennes, France

Ahmed Kamaleldin
TU Dresden
Dresden, Germany

ISSN 0302-9743 ISSN 1611-3349 (electronic)
Lecture Notes in Computer Science
ISBN 978-3-031-87896-1 ISBN 978-3-031-87897-8 (eBook)
https://doi.org/10.1007/978-3-031-87897-8

© The Editor(s) (if applicable) and The Author(s), under exclusive license to Springer Nature Switzerland AG 2025

This work is subject to copyright. All rights are solely and exclusively licensed by the Publisher, whether the whole or part of the material is concerned, specifically the rights of translation, reprinting, reuse of illustrations, recitation, broadcasting, reproduction on microfilms or in any other physical way, and transmission or information storage and retrieval, electronic adaptation, computer software, or by similar or dissimilar methodology now known or hereafter developed.
The use of general descriptive names, registered names, trademarks, service marks, etc. in this publication does not imply, even in the absence of a specific statement, that such names are exempt from the relevant protective laws and regulations and therefore free for general use.
The publisher, the authors and the editors are safe to assume that the advice and information in this book are believed to be true and accurate at the date of publication. Neither the publisher nor the authors or the editors give a warranty, expressed or implied, with respect to the material contained herein or for any errors or omissions that may have been made. The publisher remains neutral with regard to jurisdictional claims in published maps and institutional affiliations.

This Springer imprint is published by the registered company Springer Nature Switzerland AG
The registered company address is: Gewerbestrasse 11, 6330 Cham, Switzerland

If disposing of this product, please recycle the paper.

Preface

This volume contains the papers presented at the 2025 Workshop on Design and Architectures for Signal and Image Processing (DASIP 2025), which was held jointly with the 20th HiPEAC Conference in Barcelona, Spain, on January 20–22, 2025. The workshop provided an inspiring international forum for the latest innovations and developments in the fields of signal, image, and video processing and machine learning in custom embedded, edge, and cloud computing architectures and systems.

DASIP is a long-running annual workshop that was organized for the first time in 2007 in Grenoble, France, and since then it has been held in several locations in Europe and Canada, which include: Toulouse, France (DASIP 2023); Budapest, Hungary (DASIP 2022); Montreal, Canada (DASIP 2019); Porto, Portugal (DASIP 2018); Dresden and Munich, Germany (DASIP 2017 & 2024); Rennes, France (DASIP 2016); Krakow, Poland (DASIP 2015); Madrid, Spain (DASIP 2014); Cagliari, Italy (DASIP 2013); Karlsruhe, Germany (DASIP 2012); Tampere, Finland (DASIP 2011); Edinburgh, UK (DASIP 2010); Sophia Antipolis, France (DASIP 2009); and Brussels, Belgium (DASIP 2008).

For this 18th edition of DASIP, 23 papers were submitted from 9 countries around the world. Each contributed paper underwent a rigorous double-blind peer-review process during which it was reviewed by at least three reviewers who were drawn from a large pool of the Technical Program Committee members and some external reviewers. As a result, 10 high-quality papers were accepted for oral presentation at the workshop and published in these proceedings.

The success of DASIP depends on the contributions of many individuals and organizations. With that in mind, we thank all authors who submitted their work to the conference. We also wish to offer our sincere thanks to the members of the Technical Program Committee for their very detailed reviews, and to the session chairs for contributing to the success of DASIP 2025. We further extend our appreciation to the members of the Steering Committee for their support.

We address special thanks to Jose Nunez-Yanez, from Linköping University in Sweden, and to Raúl Regada Alvarez, from Thales Alenia Space (Spain), for presenting two deeply inspiring keynotes during the event, making a memorable and valuable session at DASIP 2025.

January 2025
Jordane Lorandel
Ahmed Kamaleldin

Organization

General Chairs

Jordane Lorandel — Université de Rennes, France
Ahmed Kamaleldin — Technical University of Dresden, Germany

Program Committee Chairs

Francois Berry — Institut Pascal, France
Gabriel Caffarena — CEU San Pablo University, Spain
Gustavo M. Callico — Universidad de Las Palmas de Gran Canaria, Spain
João Cardoso — University of Porto, Portugal
Daniel Chillet — IRISA/ENSSAT, University of Rennes, France
Christopher Claus — Robert Bosch GmbH, Germany
Jean Pierre David — Polytechnique Montréal, Canada
Karol Desnos — INSA de Rennes, France
Petr Dobias — CY Cergy Paris Université, France
Milos Drutarovsky — Technical University of Košice, Slovakia
Guy Gogniat — Université de Bretagne Sud, France
João Canas Ferreira — University of Porto, Portugal
Diana Göhringer — Technical University of Dresden, Germany
Marek Gorgon — AGH University of Krakow, Poland
Frank Hannig — University of Erlangen-Nurnberg, Germany
Michael Huebner — Brandenburg University of Technology, Germany
Mateusz Komorkiewicz — AGH University of Krakow, Poland
Tomasz Kryjak — AGH University of Krakow, Poland
Yannick Le Moullec — Tallinn University of Technology, Estonia
Kevin J. M. Martin — Université Bretagne Sud, France
Paolo Meloni — University of Cagliari, Italy
Arnaldo Oliviera — Universidade de Aveiro, Portugal
Andrés Otero — Universidad Politécnica de Madrid, Spain
Maxime Pelcat — INSA de Rennes, France
Sergio Pertuz — Dresden University of Technology, Germany
Christian Pilato — Politecnico di Milano, Italy
Sébastien Pillement — Nantes Université, France
Francesco Ratto — University of Cagliari, Italy

Alfonso Rodriguez   Universidad Politécnica de Madrid, Spain
Olivier Romain   CY Cergy Paris Université, France
Claudio Rubattu   University of Sassari, Italy
Ruben Salvador   CentraleSupélec, France
Yves Sorel   Inria, France
Georgios Zervakis   University of Patras, Greece

Steering Committee

João Cardoso   University of Porto, Portugal
Miguel Chavarrias   Universidad Politécnica de Madrid, Spain
Jean Pierre David   Polytechnique Montréal, Canada
Karol Desnos   INSA de Rennes, France
Diana Göhringer   Technical University of Dresden, Germany
Marek Gorgon   AGH University of Krakow, Poland
Michael Huebner   Brandenburg University of Technology, Germany
Tomasz Kryjak   AGH University of Krakow, Poland
Pierre Langlois   Polytechnique Montréal, Canada
Paolo Meloni   University of Cagliari, Italy
Sergio Pertuz   Technical University of Dresden, Germany
Andrea Pinna   Sorbonne University, France
Sébastien Pillement   Nantes Université, France
Alfonso Rodriguez   Universidad Politécnica de Madrid, Spain

Additional Reviewers

Lorenzo Casalino
Alessandro Palumbo

Invited Talks

Trends in Space Implementation of Deeptech for Signal and Data Processing

Raul Regada Alvarez

Thales Alenia Space, Spain

The space sector has been always considered one of the most innovative and top technological. However this approach relies on the difficulties and harsh environment that the satellites and spacecrafts must suffer. The massive processing is just now reaching the space sector due to the huge amount of data generated and/or directly available 'on the edge', this fact is facing a set of obstacles to overcome to properly behave under this environment. A wide set of opportunities are opening on the space sector, directly linked to the specific segment and quality approach selected, from the application of AI on the large missions where the reliability is a must to the low end segment where solutions try to mimic ground architectures but with major constraints on power or performances. The future is open to major innovations where the new products will be natively compatible with cloud computing, edge computing, AI and other deeptech. To conclude it is worth to mention that intrinsically the processing in space is clean and green once overcomed the launch process.

Hardware-Aware Adaptive Quantization in Graph Neural Networks for Inference and Training at the Edge

Jose Nunez-Yanez
Linköping University, Sweden

In this talk, we introduce a hardware approach for on-device training and inference targeting fully quantized and sparse graph neural networks (GNNs). The system supports both graph convolutional (GCN) and attention (GAT) networks with a hardware accelerator consisting of a dataflow architecture optimized for sparse data representations and adaptive fixed-point numeric precision. GATs are considered as GCNs extended with an attention mechanism that modifies the graph connectivity on-the-fly to focus attention on different parts of the input improving performance on certain applications. During training, the architecture widens the data path in the backward pass to maintain the gradient accuracy needed for back-propagation. In contrast, in the forward pass, the accelerator narrows the data path to emulate the uncertainty introduced by the quantized parameters. In the talk, we will discuss how this approach is particularly suited to reconfigurable and heterogeneous devices and the performance and energy benefits.

Contents

Specialized Hardware Architecture for Efficient Processing

CSD-Driven Speedup in RISC-V Processor 3
  Farhad Ebrahimiazandaryani and Dietmar Fey

Efficient FPGA Implementation of ViT Non-linear Functions 16
  Le Nam Hieu Nguyen and Hana Krichene

LiFT: Lightweight, FPGA-Tailored 3D Object Detection Based on LiDAR
Data 28
  Konrad Lis, Tomasz Kryjak, and Marek Gorgoń

Efficient Processing using AI for Image, Vision and Signal Applications

A Practical HW-Aware NAS Flow for AI Vision Applications
on Embedded Heterogeneous SoCs 43
  Agathe Archet, Nicolas Ventroux, Nicolas Gac, and François Orieux

Endoscopy Image Classification for Wireless Capsules with CNNs
on Microcontroller-Based Platforms 57
  Paola Busia, Andrea Pinna, and Paolo Meloni

Joint Underwater Depth Estimation and Dehazing from a Single Image
Using Attention U-Net 69
  Saqib Nazir, Reza Mohammadi Asiyabi, and Olivier Lezoray

KD-AHOSVD: Neural Network Compression via Knowledge Distillation
and Tensor Decomposition 81
  Laura Meneghetti, Edoardo Bianchi, Nicola Demo, and Gianluigi Rozza

Analysis of Emerging Techniques for Signal Processing Applications

Novel Scheduling and Shifter Networks for 5G LDPC Decoders 95
  Nikos Papageorgiou and Vassilis Paliouras

Comparison Between In-Core Hardware IDS, Off-Core Hardware IDS
and Software IDS 108
  Tianxu Li, Mohamed El-Bouazzati, Camille Monière, Philippe Tanguy,
  and Guy Gogniat

Comparative Study of Memory Optimization Techniques
for Dataflow-Modeled Applications 121
   Naouel Haggui, Maxime Pelcat, Yaesop Lee, Shuvra Bhattacharyya,
   Kevin Martin, and Jean-François Nezan

Author Index 133

Specialized Hardware Architecture
for Efficient Processing

CSD-Driven Speedup in RISC-V Processor

Farhad Ebrahimiazandaryani(✉) and Dietmar Fey

Friedrich-Alexander-Universität Erlangen-Nürnberg, Erlangen 91058, Germany
{farhad.ebrahimiazandaryani,dietmar.fey}@fau.de

Abstract. This paper introduces a synthesizable μ-architectural design method to boost the performance of a given RISC-V processor architecture by utilizing Canonical Signed Digit (CSD) representation during the execution stage within the processor pipeline. CSD is a unique ternary number system that enables carry/borrow-free addition/subtraction in constant time $O(1)$ regardless of word length N. The CSD extension is exemplarily demonstrated to the Potato processor, a simple RISC-V implementation for FPGAs. However, the method can also be applied to other implementations in principle. Our performance boost due to the CSD requires an overhead for conversion between binary and CSD representation. This overhead is compensated by an extension to a seven-stage pipeline architecture, featuring a three-step execution stage that increases the throughput and the operating frequency and enables loop unrolling, which is especially advantageous in applications with consecutive calculations, e.g., signal processing. By experimental results, we compared our CSD-based ternary solution to the original implementation, which utilizes the usual pure binary number representation of the operands. Our approach achieved a 2.41X increase in operating frequency over the original RISC-V processor on FPGA, with over 20% of this gain attributed to the CSD encoding. This enhancement resulted in up to a 2.40X improvement in throughput and a 2.37X reduction in execution time for computation-intensive benchmark applications.

Keywords: Ternary · Canonical Signed Digit · FPGA · RISC-V

1 Introduction

The ongoing demand for more computing power drives innovation in processor technology, particularly in optimizing arithmetic operations, as they directly influence overall performance. A key example is multiplication, a frequently used operation in fields like digital signal processing and machine learning, where tasks often involve consecutive multiplication operations, such as filtering. Although numerous algorithms have been developed to enhance binary multiplication, most are still limited by carry chain delays in partial product summation, reducing their efficiency. From the 1960s to the 1990s, Avizienis [1], Parhami [2] and others were already researching carry-free summation in ternary encoding that worked in $O(1)$, expediting operations like multiplication from $O((logN)^2)$ to

$O(logN)$. However, adopting this encoding comes with trade-offs: it abandons uniform number representation, introduces challenges in comparison operations, and increases memory requirements, as three states instead of two must be stored per digit. Nonetheless, CSD representation allows a unique number of representations of ternary operands and despite its potential, it has not been widely integrated into standard processor architecture. In this work, we integrate this encoding in the execution stage of a RISC-V processor and show performance increment is feasible.

This paper proposes a µ-architectural design method employing CSD representation during execution to enhance RISC-V processors' performance. Instead of using binary digits, CSD operates with balanced ternary digits $[-1, 0, +1]$, enabling carry/borrow-free operations and faster computation. The architectural shift focuses on transitioning the processor's execution stage from binary to ternary components while keeping the ISA intact. NOVA ♆, the name we gave our processor, implements the introduced method supporting RV32IZmmul_Zicsr[1] ISA, which includes integer multiplication instruction but omits hardware-based division due to its low computational frequency. This trade-off optimizes the processor for compute-intensive applications like deep learning, where multiplication is more common. The key contributions of this paper can be summarized as follows:

- **CSD Integration:** This is the first FPGA implementation of a RISC-V processor that supports "M" ISA extension using CSD-operands in its µ-Architrcture. The employed encoding features minimal Hamming weight, indicating fewer non-zero digits and reducing the number of partial products. This lowers the computation effort required for multiplication, ultimately improving performance and the latency of the execution unit in the processor.
- **Three-Step Execution:** As a technical contribution, NOVA integrates an optimized, balanced pipeline architecture in the execution stage by tripling the Instruction Execution (IE1, IE2, IE3) stage, distributing multiplication among them to increase performance. This architectural choice enables loop unrolling leading to lower branch overhead and stall cycles.

The rest of the paper is organized as follows: Sect. 2 provides an overview of Signed Digit (SD) representation, with a specific emphasis on CSD representation. Section 3 presents an extensive review of the current state-of-the-art techniques in applying SD and CSD in the digital circuit design field. Section 4 provides a comprehensive overview of the method and its implementation within an open-source RISC-V processor. Furthermore, Sect. 5 addresses the technical setup, tools, and benchmarks where NOVA is deployed, along with a detailed presentation of the FPGA-based evaluation results. Finally, Sect. 6 concludes the paper by summarizing the key findings and contributions of this research work.

---

[1] The RV32IZmmul_Zicsr ISA provides a RISC-V processor with a base set of integer arithmetic instructions (RV32I), Multiplication extension excluding division (Zmmul), and instructions for manipulating control and status registers (Zicsr).

2 Preliminary

In this section, a concise overview of the key concepts associated with SD representation, specifically the CSD representation and its properties, is presented as follows.

2.1 SD Representation

Researchers have investigated the utilization of SD representation as an alternative approach to address the limitations associated with the conventional binary representation [1,2]. According to a definition in [2], a generalized SD number system to a radix r utilizes a digit set $S = \{-\alpha, -(\alpha - 1), \ldots, -1, 0, 1, \ldots, \beta\}$, with $\alpha, \beta > 0$ and $\alpha + \beta + 1 > r$ and $r > 1$. E.g., for the so-called balanced Binary Signed Digit (BSD) system holds, that the digit set is symmetric ($\alpha = \beta = 1$), and for the radix holds, $r = 2$, which yields the digit set $S = \{-1, 0, +1\}$. SD representation's carry/borrow-free addition/subtraction property accelerates arithmetic operations and facilitates efficient computation in various applications [2]. Figure 1 illustrates the detailed design of a 32-digit adder/subtractor for two provided SD numbers[2], $x = x_{31}^n x_{31}^p x_{30}^n x_{30}^p \ldots x_1^n x_1^p x_0^n x_0^p$ and $y = y_{31}^n y_{31}^p y_{30}^n y_{30}^p \ldots y_1^n y_1^p y_0^n y_0^p$. Notably, the critical path has been constrained to two adjacent digits, regardless of the input data word length. Nonetheless, regular SD representation has drawbacks: it lacks uniqueness, allowing multiple representations of the same number that complicates comparisons. Additionally, it possesses a non-minimal Hamming weight, leading to more effort in result generation (e.g., more partial products in multiplication), and reducing performance improvements.

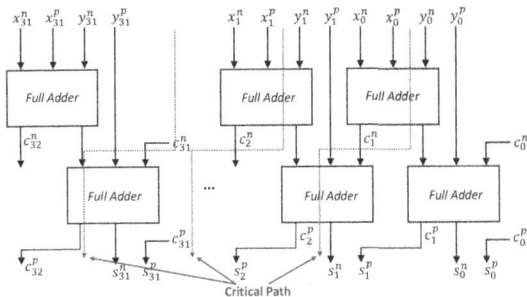

Fig. 1. 32-digit SD/CSD Adder/Subtractor

---

[2] $x = (x_i^n x_i^p)$ is to interpret as follows. In a strict digital system a ternary digit requires two binary values, e.g. x_i^n, and x_i^p, called plus-minus coding. It codes a ternary digit x as follows $x = x_i^p - x_i^n$, e.g. $(1,0)$ is equal to -1.

2.2 CSD Representation

CSD representation is a specific form of radix-2 SD representation that possesses distinct properties as follows [3]:

- Adjacent CSD digits are never both non-zero, meaning $|d_i \cdot d_{i-1}| \neq 1$, where $1 \leq i \leq N-1$, and N is the number of digits.
- It exhibits the minimum number of non-zero digits among various representations. Consequently, the probability of a CSD digit denoted as d_i being non-zero is determined by:

$$P(|d_i| = 1) = \frac{1}{3}\left(1 + \frac{1}{4N}\right)$$

- Utilizing the first property guarantees that each arbitrary number has a unique and exclusive encoding ensuring the distinct representation of numbers.

The first two properties of CSD representation show that as the number of digits N increases, the probability of a non-zero digit approaches $\frac{1}{3}$, reducing the Hamming weight and capping the number of non-zero partial products to $\frac{N}{2}$ in multiplications. In contrast, binary or regular SD representations have a fixed probability of $\frac{1}{2}$, potentially producing up to N non-zero partial products. Additionally, the third property allows for simple, digit-wise comparisons within the ternary encoding, addressing the non-uniqueness issue present in regular SD representation (see [4]).

2.3 Encoding

A radix-2 CSD representation is a redundant expression of binary numbers that allows for more than 2 values per digit. In other words, each digit d_i should be encoded using two bits ($r = 2$). In practice, Table 1 showcases the two most commonly utilized encodings [5]. Encoding 1 can be characterized as a sign-value representation that adheres to the following relation:

$$d_i = (-1)^{d_i^s} \times d_i^v \tag{1}$$

where d_i^s and d_i^v represent the sign bit and the value bit, respectively. To determine the encoding that corresponds to each digit in the first scheme, Eq. (1) provides the corresponding computation. Encoding 2 called Positive-Negative, allows an additional valid representation of 0 when $d_i^p = 1$ and $d_i^n = 1$, which is useful in some arithmetic implementations and is used in this research. The corresponding encoding of each digit in the second scheme can be achieved by employing Eq. (2).

$$d_i = d_i^p - d_i^n \tag{2}$$

Table 1. Two common encodings used in the binary representation of a CSD digit

SD/CSD digit	Encoding 1		Encoding 2	
d_i	d_i^s	d_i^v	d_i^p	d_i^n
−1	1	1	0	1
0	0	0	0	0
0	1	0	1	1
1	0	1	1	0

3 Prior Works

This section thoroughly explores the latest techniques and designs related to SD and CSD representation. By examining these approaches, it aims to provide a detailed understanding of the current advancements in utilizing these representations.

Recent research [4] has shown that switching from conventional binary number representation to alternative schemes can significantly improve processor arithmetic logic unit performance. This is due to the reduced logic depth in these systems, which can surpass the speed of traditional binary circuits. However, a key limitation is the lack of unique representations for integers, requiring the processor to revert to binary for tasks like comparisons, where distinct number representation is necessary.

The work in [6] focuses on a ternary processor leveraging Carbon Nanotube Field Effect Transistor (CNTFET)-based ternary logic for the control unit and memory. This approach demonstrates performance improvement and power efficiency compared to traditional binary systems. However, it faces challenges at the circuit level, particularly with carry chain propagation. Unlike our method, which uses base-2 with CMOS technology and benefits from established infrastructure, their design employs a base-3 system and does not adopt a redundant number representation, complicating effective carry chain management and limiting optimization potential for processor performance.

The work presented in [7] introduces a comprehensive design and verification framework for developing a fully functional emerging ternary processor utilizing balanced ternary number representation based on ternary logic. While arithmetic operations like addition and subtraction show improved efficiency in ternary logic, more complex operations such as multiplication present additional challenges compared to binary counterparts. The results primarily remain conceptual, lacking concrete implementation details. In contrast, our approach is grounded in a binary logic system that operates on ternary-represented numbers as data operands. In the work by [8], CSD encoding has been leveraged in the development of a Neural Network-based image compression architecture. While effectively deployed on FPGA platforms for grayscale image processing, the investigation underscores the unexplored domain of extending this methodology to color image compression, attributed to the intricate complexities associated with the employed encoding scheme.

Researchers in [9,10] have explored the use of CSD-encoded coefficients in FIR filter designs, highlighting several key benefits. Specifically, CSD reduces the number of additions required during multiplication, a central operation in FIR filters, resulting in lower computational complexity and improved numerical stability, especially in FPGA implementations. However, while non-binary digits boost filter performance, they can introduce complexity and increase implementation costs, which is particularly problematic for smaller-sized FIR filters where the overhead might outweigh the performance gains. These insights into CSD's efficiency in filter design have inspired its application in the execution stage of general-purpose processors, aiming to tackle computation bottlenecks effectively.

In [11], the authors introduced a ternary representation to improve processor performance and computational efficiency within their proposed architecture. However, this design is constrained by its support for only the base RV32I instruction set, limiting its applicability in AI and signal processing fields where performing additional instructions directly can offer more speed-up compared to the traditional indirect techniques e.g. add/shift loop for multiplication.

While researchers have explored the application of SD/CSD representation across various domains, yielding valuable outcomes, none of the proposed solutions specifically addressed the demands of computation-intensive applications to alleviate bottleneck issues related to binary encoding in processors.

4 Proposed Design Method in a RISC-V Core

This section presents an in-depth overview of the proposed method and its implementation in an open-source RISC-V processor [12] that suffers from a computation bottleneck in the execution stage.

To harness CSD computation benefits initially, integrating a Binary-to-CSD (B2C) circuit is unavoidable. This circuit converts incoming binary values into CSD numbers in IE1, before their utilization in the next steps for arithmetic instructions execution. We are favoring [5] due to its low time complexity which is employed in NOVA architecture for B2C purposes. Refer to [5] for detailed conversion guidelines. The efficient computation achieved by CSD representation is attributed to the constrained carry chain (see Fig. 1), limited to two adjacent digits. Subtraction operations benefit from this constraint as the requirement for the negative value of the second operand can be met by individually interchanging the positive and negative bits (y_i^p and y_i^n) for each digit. This approach allows the adder circuit to effectively compute $A + (-B)$ instead of requiring a separate circuit for $A - B$. Consequently, the overall circuit design is simplified, and the computational efficiency of subtraction operations is enhanced. In binary representation, $N \times N$ multiplication produces N non-zero partial products. In contrast, CSD multiplication reduces this number to $N/2$ in the worst case because at least one of every two adjacent digits is always zero, resulting in a zero partial product. A bitwise OR operation on adjacent partial products, as

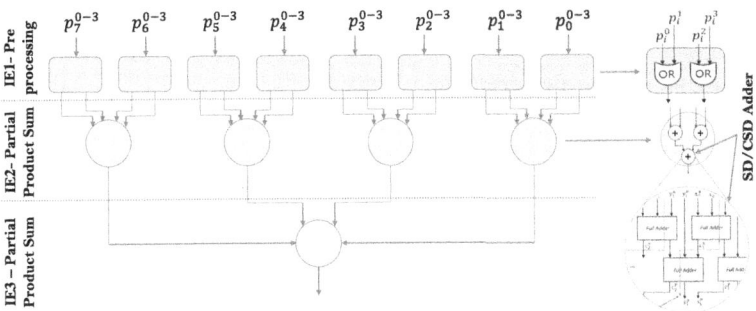

Fig. 2. Pipelined CSD multiplier unit along with internal structure of employed blocks.

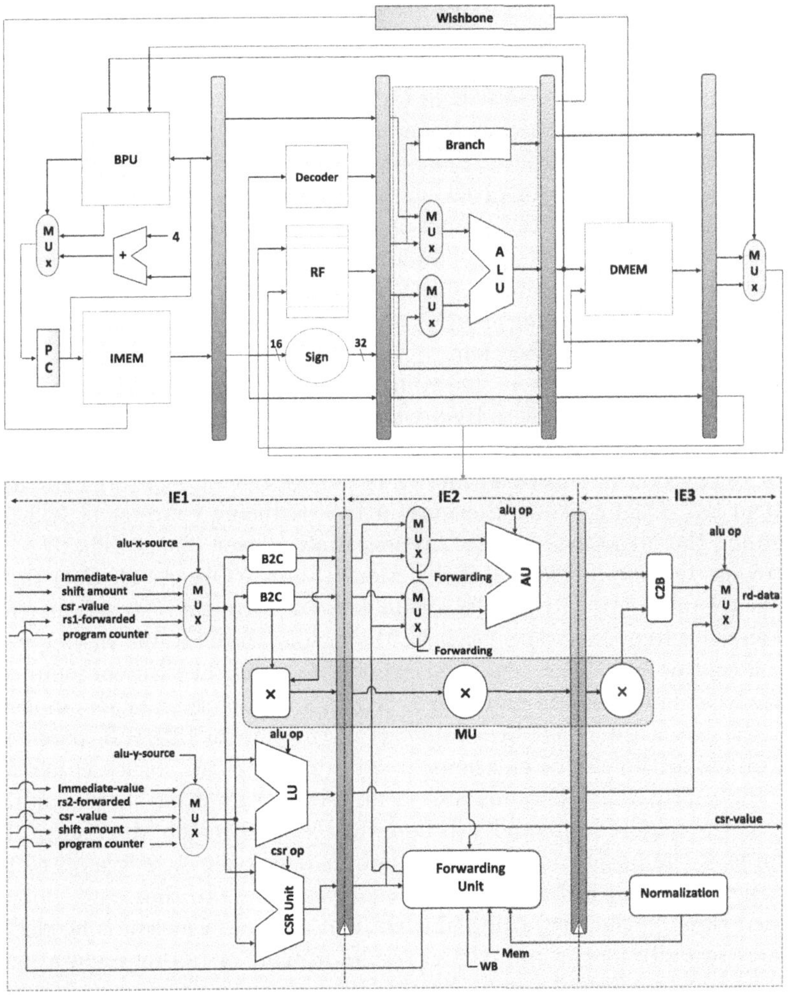

Fig. 3. The overall architecture of NOVA with a detail view of the execution stage.

shown in Fig. 2, effectively filters out the zero values. This allows only the non-zero products to continue to the next stages of the multiplication tree in IE2 and IE3, enhancing overall performance. Additionally, using SD/CSD adder circuit for partial sum computation improves performance from $O(\log N)^2$ to $O(\log N)$, thereby increasing the efficiency of arithmetic operations. After the execution unit performs the instruction and generates the final result, it becomes necessary to convert the result back into binary representation for storage in memory or register files. The C2B (CSD to binary) conversion is straightforward and can be achieved by decomposing the result into a positive (y^p) and one negative (y^n) vector. This is followed by performing binary summation of y^p and $y^{n'}$ with $C_0 = 1$, where $y^{n'}$ is the one's complement representation of y^n. The output is the binary equivalent of the CSD unit result, excluding the overflow bit in this sequence. All these steps can be executed in $O(\log N)$.

Certainly, the developed arithmetic instructions involve several steps. Initially, the input values are transformed from binary to CSD using the B2C module. Subsequently, the Arithmetic Unit (AU)/Multiplication Unit (MU) executes the instruction using the converted operands. Finally, the outcome is converted to binary encoding using the C2B module, yielding the final result. However, the sequential conversions involved in this process may introduce delays that impact the critical path of the overall operation, potentially resulting in slower performance compared to binary computations and diminishing the advantages of fixed time-frame calculations using CSD. Increasing the number of pipeline stages emerges as a promising solution to address this bottleneck, which is discussed in detail in the next section.

The architecture of NOVA, shown in Fig. 3, consists of the IF, ID, IE (IE1, IE2, IE3), Mem, and WB stages. By dividing the execution stage into three parts, the B2C conversion result can be accessed immediately by the MU for partial product calculation in constant time $O(1)$ during IE1 or registered for use by the AU in IE2. C2B conversion occurs in IE3, reducing conversion delays and maintaining the advantages of CSD representation over binary. The IE1 stage also provides outputs for logical instructions (Shift, XOR, AND, OR) through the Logical Unit (LU), with results available for subsequent instructions via the forwarding unit to prevent data hazards. Meanwhile, the CSR unit manages operations on control and status registers, essential for tasks like system monitoring. Data hazards require careful handling in NOVA's architecture. When an instruction in IE1 needs the result of a previous instruction in IE2/IE3, particularly one involving arithmetic calculations, a stall of one or two cycles is necessary. This stall ensures that operands are accurately retrieved for the computation in IE1, as the required operand will be available later in the Mem stage. However, suppose the output from an Add/Sub instruction in IE2 is used as input for another Add/Sub instruction in subsequent cycles. In that case, it can be forwarded through the forwarding unit without stalling. Caution is needed here because repeatedly forwarding partial results in CSD or SD representation can lead to pseudo-overflow. To mitigate this issue, a normalization unit is essential to address the problem in a fixed time without significantly affecting circuit performance. As discussed earlier, due to the deeply pipelined execution unit, NOVA can support loop unrolling for both arithmetic and logical instructions. This reduces stall cycles and efficiently fills the 3-step pipeline of the execution stage, resulting in higher throughput. It is important to note that logical operations, such as Shift, AND, OR, and XOR, must be executed using traditional binary representation. This limitation arises from CSD encoding, which primarily optimizes arithmetic instructions but does not enhance logical operations. However, since these logical instructions are not part of the critical path, they do not disrupt the overall functionality of the data path. Consequently, while a different approach is necessary for these operations, their effect on the circuit's performance and data path functionality is minimal.

5 Result and Discussion

This section details the technical setup and benchmarks used for the FPGA experiments. It presents the results obtained from an open-source RISC-V core [12], the Potato processor (PRISC-V), as the original core tested with various configurations, alongside results from NOVA.

5.1 Technical Setup, Tools, and Benchmarks

In FPGA implementation, the Xilinx Vivado 2023.1 design suite software was used for synthesis, mapping, placement, and routing. Additionally, the Ultra96-v2 (XCZU3EG-1SBVA484I) from the Zynq Ultrascale+ development board family was chosen as the FPGA platform to establish results in all tables and figures in this section. The results provided are specifically derived from the analysis conducted using the specified synthesis tool and FPGA board chosen for the processor's implementation and may differ when using alternative synthesis tools or FPGA boards, reflecting variations in architecture and optimization strategies. The efficiency evaluation of cores made use of applications from the Mibench suite [13], as well as three well-known applications widely used for processor performance, memory bandwidth examination, and matrix multiplication, namely Dhrystone [14], Schönauer vector triad [15], and Matmul [16]. The chosen applications span various domains, including signal processing and telecommunications, automotive, networking, error-checking, sorting, and arithmetic calculations.

We conducted a thorough evaluation of performance metrics, including throughput, execution time, and resource utilization, to assess computational efficiency in NOVA versus PRISC-V(full binary). Both support the same 32-bit ISA and in-order execution. PRISC-V configurations include a 5-stage pipeline with/out DSP (PRISC-V$_{DSP}$/PRISC-V), a 7-stage pipeline with/out DSP (PRISC-VII$_{DSP}$/PRISC-VII) featuring a 3-step execution stage. All five architectures are identical as Fig. 3 except IE stage where PRISC-V configurations do not include conversion circuits and its 7-stage configurations the multiplier is implemented in three steps: IE1 computes four partial products in parallel ($PP_1 = X_{15:0} \times Y_{15:0}$, $PP_2 = X_{31:16} \times Y_{15:0}$, $PP_3 = X_{15:0} \times Y_{31:16}$, and $PP_4 = X_{31:16} \times Y_{31:16}$). IE2 and IE3 add these partial products in a binary tree model to produce the final result.

Table 2. Comparative Logic and Network delay for binary/CSD multipliers

Article	Net delay (ns)	Logic delay (ns)
BinMul$_{DSP}$	3.441	4.201
BinMul	8.132	3.027
CSDMul	8.305	2.256

5.2 Experimantal Results

Initially, we analyzed the critical path delays of a 32-bit binary ALU and a 32-digit CSD ALU, both passing through the multiplier module. Table 2 illustrates that the binary multiplier, utilizing DSP (BinMul$_{DSP}$), features a logic delay of 4.201 ns and a network delay of 3.441 ns. In contrast, the binary multiplier without DSP utilization (BinMul) has a lower logic delay of 3.027 ns but suffers from a higher network delay of 8.132 ns, reducing the efficiency of a non-DSP binary multiplier. Despite an optimized logic delay of 2.256 ns in the CSD multiplier, it has a substantial network delay of 8.305 ns, overshadowing its performance. A more detailed analysis shows that 35% and 48% of the reported logic and network delays in CSD multiplication are due to the conversion circuits. Deeply pipelining the multiplication unit effectively reduces network delay, making designs with lower logic delays well-suited for enhancing performance. As a result, the CSD multiplier stands out as a compelling option.

The frequency values reported in Table 3 represent the maximum frequencies that each core was able to operate with, ensuring there were no timing violations or clocking issues. As depicted in Table 3, NOVA achieves a 2.41X frequency boost over PRISC-V, with a 1.62X increase in FFs and 1.37X in LUTs due to its 3-step execution and encoding scheme. When compared to PRISC-V$_{DSP}$, NOVA improves frequency by 2.12X which comes with an increase in resource usage-2.61X more LUTs and 1.85X more FFs while PRISC-V$_{DSP}$, uses 12 DSP

Table 3. Resource Utilization and Maximum Operating Frequency

Article	SRC (LUTs, FFs, DSP)	Freq_Max (MHz)
PRISC-V	4212, 1425, 0	114
PRISC-V$_{DSP}$	2206, 1251, 12	130
PRISC-VII	4641, 2141, 0	225
PRISC-VII$_{DSP}$	2660, 1864, 4	225
NOVA	5762, 2309, 0	275

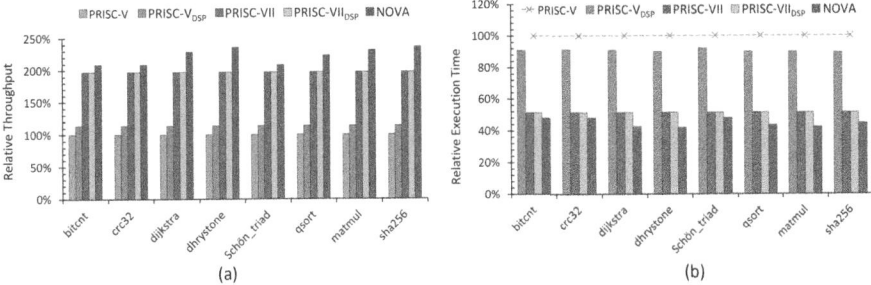

Fig. 4. Relative Throughput and Execution time for Various Applications.

blocks occupying a large area. Deep pipelining enhances PRISC-VII's frequency by 1.97X, matching PRISC-VII$_{DSP}$'s, 1.74X improvement. NOVA, achieves 22% better performance, though it consumes 1.23X LUTs and 1.08X FFs than PRISC-VII, and 1.24X FFs than PRISC-VII$_{DSP}$. Despite higher resource usage, NOVA remains competitive due to its frequency gains. In Fig. 4 (a) and Fig. 4 (b), we compared NOVA with all PRISC-Vconfigurations to illustrate the performance improvement achieved by the employed method in terms of throughput and execution time. These metrics are measured using internal timers. The results have been normalized to the outcomes of the 5-stage DSP-less design, PRISC-V, serving as a reference for evaluating the efficiency of other cores. NOVA achieves a 2.40X throughput improvement over PRISC-V, especially in compute-intensive tasks like Dhrystone, SHA256, and Matmul, and 2.17X in memory-bound applications like Schönauer vector triad and CRC32. When compared to PRISC-V$_{DSP}$, NOVA shows a 2.10X boost in compute-intensive tasks and 1.83X in less demanding ones. The increase in pipeline stages for both PRISC-V and PRISC-V$_{DSP}$ yields a 1.96X and 1.73X improvement respectively, but NOVA still surpasses them by over 22%, thanks to its efficient encoding scheme. Regarding execution time, NOVA offers a 2.37X reduction in compute-intensive tasks and a 2.11X reduction for memory-bound tasks over PRISC-V. PRISC-VII and PRISC-VII$_{DSP}$ achieve 1.95X and 1.73X reductions compared to their 5-stage counterparts. However, NOVA still achieves over 20% better performance than both of them for compute-intensive tasks highlighting the effectiveness of NOVA's CSD encoding and its capability to enhance processor performance. To evaluate NOVA's loop-unrolling capability, we tested it on the Schönauer vector triad, a memory-bound application. Despite inherent performance limitations, 3-way unrolling for 9×10^6 iterations reduced execution time from 3.5 s to 2.5 s. However, due to limited compiler support and insufficient instructions, the full theoretical benefits were not achieved.

6 Conclusion

In conclusion, this paper presented a synthesizable μ-architectural design method that improved RISC-V processor performance by integrating CSD representation into the execution stage and extending the pipeline to seven stages. While the method introduced some conversion circuits and resource utilization overheads, it proved to be effective, especially in computation-hungry applications like signal processing. Its implementation on an open-source RISC-V processor showcased its adaptability to other architectures, making it a promising solution for performance enhancements. Experimental results demonstrated a 2.41X boost in frequency, up to a 2.40X increase in throughput, and a 2.37X reduction in execution time for computation-intensive benchmark applications, where over 20% of the gain was due to the CSD encoding.

References

1. Avizienis, A.: Signed-digit number representations for fast parallel arithmetic. IRE Trans. Electron. Comput. **3**, 389–400 (1961)
2. Parhami, B.: Generalized signed-digit number systems: a unifying framework for redundant number representations. IEEE Trans. Comput. **39**(1), 89–98 (1990)
3. Hashemian, R.: A new method for conversion of a 2's complement to canonic signed digit number system and its representation. In: Conference Record of the Thirtieth Asilomar Conference on Signals, Systems and Computers. IEEE (1996)
4. Reichenbach, M., Knödtel, J., Rachuj, S., Fey, D.: RISC-V3: a RISC-V compatible CPU with a data path based on redundant number systems. IEEE Access **9**, 43684–43700 (2021)
5. Ruiz, G.A., Granda, M.: Efficient canonic signed digit recoding. Microelectron. J. **42**(9), 1090–1097 (2011)
6. Karthikeyan, S., Reddy, M.C.K., Monica, P.R.: Design of CNTFET-Based ternary control unit and memory for a ternary processor. In: 2017 International Conference on Microelectronic Devices, Circuits, and Systems (ICMDCS), pp. 1–4. IEEE (2017)
7. Kam, D., et al.: Design and evaluation frameworks for advanced riscbased ternary processor. In: 2022 Design, Automation Test in Europe Conference Exhibition (DATE). IEEE (2022)
8. Kiran, M.L., Nikhileswar, K., Ramanaiah, K.V.: FPGA implementation of CSD-based NN image compression architecture. ICTACT J. Microelectron. **16**(102), 102 (2021)
9. Srivastava, A.K., Raj, K.: An efficient FIR filter based on hardware sharing architecture using CSD coefficient grouping for wireless applications. Wireless Pers. Commun. **123**, 3433–3448 (2022)
10. Lee, H., Sobelman, G.E.: FPGA-based digit-serial CSD FIR filter for image signal format conversion. Microelectron. J. **33**(5–6), 501–508 (2002)
11. Azandaryani, F.E., Fey, D.: ExTern: boosting RISC-V core performance using ternary encoding, Microprocessors and Microsystems. 107 (2024). https://doi.org/10.1016/j.micpro.2024.105058
12. Skordal, K.K, Siorpaes, D., Cotret, P., Thomas, J.: Potato Project (2023). https://opencores.org/projects/potato

13. Guthaus, M.R., et al.: MiBench: a free, commercially representative embedded benchmark suite. In: Proceedings of the Fourth Annual IEEE International Workshop on Workload Characterization. WWC-4 (Cat. No. 01EX538). IEEE (2001)
14. Weicker, R.P.: Dhrystone: a synthetic systems programming benchmark. Commun. ACM **27**(10), 1013–1030 (1984)
15. Hofmann, J., Eitzinger, J., Fey, D.: Execution-cache-memory performance model: introduction and validation (2015). arXiv preprint arXiv:1509.03118
16. Benson, A.R., Ballard, G.: A framework for practical parallel fast matrix multiplication. ACM SIGPLAN Notices **50**(8), 42–53 (2015)

Efficient FPGA Implementation of ViT Non-linear Functions

Le Nam Hieu Nguyen and Hana Krichene(✉)

CEA, List, Palaiseau 91120, France
{le-namhieu.nguyen,hana.krichene}@cea.fr

Abstract. Building on the performance of Transformers, Vision Transformers (ViTs) have recently been applied across a wide range of Computer Vision (CV) tasks, demonstrating superior results compared to traditional Convolutional Neural Networks (CNNs). However, the computational demands of ViTs remain a concern when deployed on edge devices, primarily due to the complexity of their various layers. Unlike CNNs, ViTs rely on multiple non-linear functions (Softmax, GELU, and LayerNorm) that significantly increase resource utilization, power consumption, and potentially induce high latency. Addressing the lack of FPGA-based implementations of non-linear functions, this work proposes a resource-efficient FPGA solution that achieves a 5.8× reduction in Look-Up Tables (LUTs) and a 12.3× reduction in registers, with a throughput of 32 processed elements per cycle and a clock frequency of 200 MHz, enabled by a reuse technique. Additionally, latency is reduced by 1.7× through a proposed three-stage pipelined architecture. The design is implemented on the Xilinx XCVU9P FPGA, leveraging decomposition techniques to resolve data dependencies and analyze similarities among the three non-linear functions in ViTs with the aim of combining them into a unified solution. A second-order mathematical approximation is employed to facilitate efficient synthesis of these non-linear functions on the FPGA.

Keywords: Non-linear Functions · Vision Transformer · Resource reuse · FPGA

1 Introduction

In the last five years, a new deep learning architecture called Transformer has emerged as one of the most efficient methods for processing data, gradually replacing earlier neural network architectures [1]. Inspired by the success of the Attention mechanism in the Transformer for natural language processing, ViT was developed by applying Attention in the field of CV. However, when it comes to inference on edge devices, implementing ViT is a challenging task due to the enormous computational complexity and resource requirements. In particular, the ViT architecture includes several non-linear layers, which make its deployment on embedded devices difficult due to limited computational resources. While non-linear functions in CNNs can be replaced with more hardware-friendly alternatives (e.g. ReLU, LeakyReLU), or some (like Softmax) can be omitted since they are only used in the final layer for probability calculation, the non-linear functions in ViT are integral to the inference process and are present across multiple layers. Moreover, although many AI accelerators have been developed, most are specialized for CNNs, which have dominated the AI field for a long time. Consequently, there is a clear and growing need for research, innovation, and the implementation of AI accelerators specifically adapted to ViTs.

Several FPGA-based accelerators have been proposed to exploit the ViTs, leveraging the flexibility and scalability of FPGAs. However, most of these accelerators focus predominantly on the linear parts, as they represent the major portion of the computations. In reality, ViT computations alternate between linear matrix multiplications and non-linear layers (LayerNorm, Softmax, and GELU). Neglecting the non-linear parts can result in inefficient switching during the inference process, leading to suboptimal performance. There are three main bottlenecks in the hardware implementation of non-linear functions in ViTs. First, non-linear operations (such as division, exponentiation, square root, etc.) are computationally expensive on edge devices. Second, the three data-dependent loops within LayerNorm and Softmax introduce considerable latency. Lastly, resource limitations pose a significant challenge for implementing these non-linear functions. Although non-linear functions occur frequently in ViTs, their computational contribution is minor compared to the linear parts. Allocating substantial resources for non-linear implementations could lead to inefficient resource utilization and negatively impact the implementation of the linear components.

In this paper, we do not focus on the three non-linear functions in ViT individually. Instead, we conduct an analysis of the similarities among these functions, which allows us to minimize the number of computational units. Our main contributions are as follows: (1) We simplify the non-linear operations using second-order approximation, enabling the datapath to consist of simple operations such as shifters, adders, and multipliers. (2) We propose a decomposition technique to break down the three functions into basic computational units, highlighting their similarities and enabling shared resource utilization across the functions. (3) Based on these simpler computational units, a three-stage pipeline is employed to process data simultaneously, thereby mitigating high latency caused by data dependencies.

The remainder of this paper is organized as follows: Sect. 2 presents the background and related work. Section 3 introduces the proposed architecture. The experimental results, including accuracy and synthesis, are discussed in Sect. 4. Finally, Sect. 5 concludes the paper.

2 Background

2.1 Non-linear Functions In Vision Transformer

Softmax. The exponential function can potentially lead to overflow and cause accuracy loss. Therefore, it is usually computed in the negative domain. Consequently, the Softmax function in ViT can be expressed as follows:

$$\text{Softmax}(x_i) = \frac{e^{x_i - x_{max}}}{\sum_{j=1}^{n} e^{x_j - x_{max}}} = \frac{2^{\log_2 e(x_i - x_{max})}}{\sum_{j=1}^{n} 2^{\log_2 e(x_j - x_{max})}} \quad (1)$$

where $x_0, x_1, \ldots, x_{n-1}$ and x_{max} are the Softmax inputs and the maximum value among those inputs, respectively. Noticeably, using the natural exponential function e is not hardware-friendly, so base e is often replaced with base 2 to optimize resource usage.

The Softmax function has two data dependencies, which means that two passes through all inputs are required before the first output can be generated: one pass for finding the maximum value and another for accumulating the values for the denominator. Consequently, this constraint can become a bottleneck in processing if the design is not carefully considered, due to the time spent waiting for all necessary inputs.

LayerNorm. The normalization function also has two data dependencies, similar to Softmax. Essentially, all inputs are processed to calculate the average value (mean). A second pass is then conducted to compute the variance (the mean square value of the inputs minus the mean) before calculating the first normalized output. The formula for LayerNorm is expressed as follows:

$$\text{LayerNorm}(x_i) = \frac{x_i - E(x)}{\sqrt{\text{Var}(x)}} \gamma + \beta \quad (2)$$

where $x_0, x_1, \ldots, x_{n-1}$, $E(x)$, and $\text{Var}(x)$ are respectively the inputs, mean, and variance of the inputs.

GELU. Different from Softmax and LayerNorm, the GELU function operates data-independently. However, the complexity of this function is a true hindrance in hardware design, as shown:

$$\text{GELU}(x) = x \cdot \frac{1}{2}\left[1 + \frac{2}{\sqrt{\pi}} \int_{0}^{\frac{x}{\sqrt{2}}} e^{-t^2} dt\right] = x \cdot L\left(\frac{x}{\sqrt{2}}\right) \quad (3)$$

2.2 Second Order Approximation

There are multiple works focused on approximating non-linear operations in Transformers to achieve high accuracy [2–4]. However, most of these studies primarily concentrate on approximating the natural exponential function used in Softmax and the distribution function $L(x)$ in GELU, while other non-linear operations such as division or square root are often overlooked. Furthermore, the methods employed for approximation are usually specific to individual operations, leading to inconsistencies in accelerator design and limiting resource reuse among non-linear operations, as each operation requires its own dedicated component. Building on the excellent accuracy results demonstrated in [2], this work leverages the power of second-order polynomial approximation to address all non-linear operations in the accelerator. This approach includes approximating the exponential function, square root, distribution function $L(x)$, and division, with the aim of unifying the architecture of all computational components and enhancing resource efficiency.

Given a set of data $\{(x_0, y_0), (x_1, y_1), \ldots, (x_n, y_n)\}$ where $n \geq 2$ and $y = f(x)$ is a non-linear function, the second-order approximation of this non-linear function $f_{2\text{nd}}(x)$ can be inferred using the Least Squares method by solving the following problem:

$$\min_{a,b,c} \sum_{i=0}^{n} (y_i - f_{2\text{nd}}(a, b, c, x_i))^2 \qquad (4)$$

where $f_{2\text{nd}}(x) = a(x+b)^2 + c$ or $f_{2\text{nd}}(x) = a[(x+b)^2 + c]$.

2.3 Related Work

Parallel to the development of ViT, several studies [3–7] focus more on non-linear functions optimization in ViT. These solutions employ various approximation methods aimed at reducing area and enhancing throughput. While some proposals concentrate solely on optimizing Softmax, others address all types of non-linear functions within Transformers. Study [5] replaces Softmax with a hardware-friendly similarity measurement method, whereas [7] delves into optimizing the implementation of this function on FPGA. Meanwhile, study [4] avoids the use of any DSP by employing shift operations, resulting in minimal area overhead and high frequency. Although it represents the state-of-the-art in resource utilization by reusing the exponential function in GELU and Softmax, there is no optimization for LayerNorm. Study [6] applies piece-wise approximation, mapping each segment of non-linear functions in the BERT model into multiple linear segments. This approach can minimize resources due to operation reuse (addition and comparison), but the variable and broad range of non-linear functions may require a large amount of LUTs to store mapping parameters and can lead to accuracy loss. Study [3] inherits techniques from previous studies and proposes an architecture that effectively leverages parallelism and pipelining to achieve negligible latency.

All of the above accelerators focus primarily on dataflow schemes, or only partially investigate non-linear functions in the ViT model on FPGA. Although several works, such as [8,9], consider the similarities of Softmax, LayerNorm, and GELU to create a specialized computational unit that reuses resources for all three functions, they are oriented towards ASIC technology or 32-bit processing, and they lack flexibility for integration into different Matrix Multiplication acceleration cores. Moreover, similar to [6], even though [8] leverages neural network algorithms to more accurately determine parameters for piece-wise functions, the wide range of inputs can lead to significant storage requirements. Instead, our work prioritizes resource reuse through a second-order polynomial approximation. In this approach, all inputs are carefully scaled down to a specific range, making computation feasible within this range by using operations suitable for only that range. Additionally, a three-stage pipeline is implemented to address the latency problem caused by data dependence in Softmax and LayerNorm.

3 Proposed Architecture

3.1 Three-Stage Pipelined Architecture

A first challenge in processing non-linear functions in ViT is data dependency. In our implementation, we propose pipelining these functions at a higher level of abstraction before applying pipelining at the operational level. Specifically, the functions are decomposed into smaller components based on their data dependencies. This approach allows dependent data in different groups to be processed concurrently, significantly improving latency efficiency.

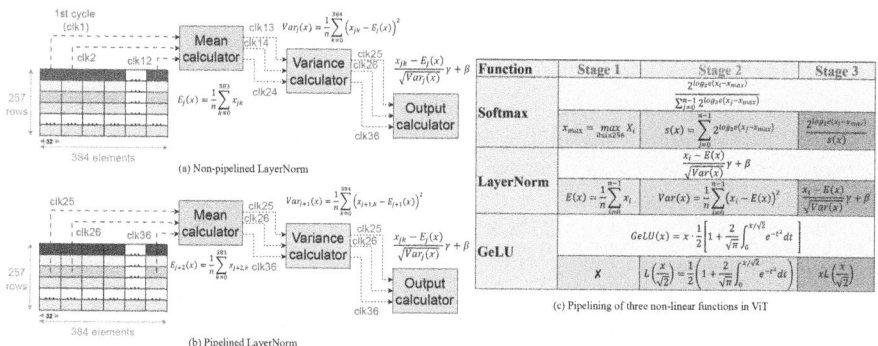

Fig. 1. Latency reduction through concurrent processing of pipeline architecture and decomposition of nonlinear functions.

Figure 1(a) illustrates the processing time for LayerNorm with an input matrix sized $(257, 384)$ without any optimization. LayerNorm operates under a row-wise constraint, requiring all data in one row of the input (384 elements) to begin computation. With a throughput of 32, it takes 12 cycles to generate the average value of the first row. Then, the variance of the first row is calculated over another 12 cycles, and an additional 12 cycles are needed to complete the processing of one row. With a latency of 36 cycles per row, the total time required to process the input matrix $(257, 384)$ is 9252 cycles.

On the other hand, by applying the pipelining technique through implementing the calculator blocks independently, the processing time of the input matrix can be reduced by a factor of 3. In Fig. 1(b), each calculator operates simultaneously, meaning that the Mean calculator and Variance calculator do not remain idle while the Output calculator has not yet finished. Instead, these calculators continue processing the next row, enabling concurrent processing of up to 3 rows. As shown, at the 36th cycle, while a non-pipelined implementation has only completed the processing of the first row and has not yet started on the subsequent rows, the pipelined one has already completed the calculation of the second row's variance and the third row's mean. Although the latency per row remains at 36 cycles, the total latency for the input matrix is reduced to 3108 cycles ($36 + 256 \times 12$), representing nearly a 3 times acceleration.

3.2 Resource Reuse

A second challenge in the implementation of non-linear functions is the resource constraint. To minimize computational resource usage, these functions can be analyzed at a lower level of abstraction, where their similarities are highlighted. This enables the application of a reuse technique, allowing common computational units to be shared among the functions.

Function	Stage 1	Stage 2					Stage 3	
Softmax	$x_{max} = \max\limits_{0 \le i \le 256} X_i$	$x'_i = x_i - x_{max}$	$x''_i = \log_2 e \cdot x'_i$	$p_i = 2^{x''_i}$	$s(x) = \sum\limits_{i=0}^{256} p_i$ $\frac{2^{\log_2 e(x_i - x_{max})}}{\sum_{j=0}^{n-1} 2^{\log_2 e(x_j - x_{max})}}$	$I(x) = \frac{1}{s(x)}$	$y_i = I(x)p_i$	X
LayerNorm	$E(x) = \frac{1}{n}\sum\limits_{i=0}^{384} x_i$	$x'_i = x_i - E(X)$	$x''_i = (x'_i)^2$	X	$Var(x) = \frac{1}{n}\sum\limits_{i=0}^{n-1} x''_i$ $\frac{x_i - E(x)}{\sqrt{Var(x)}}\gamma + \beta$	$I(x) = \frac{\gamma}{\sqrt{Var(x)}}$	$y'_i = I(x)x'_i$	$y_i = y'_i + \beta$
GeLU	X	X	$x' = \frac{1}{\sqrt{2}}x$ $GeLU(x) = x \cdot \frac{1}{2}\left[1 + \frac{2}{\sqrt{\pi}}\int_0^{x/\sqrt{2}} e^{-t^2} dt\right] = xL\left(\frac{x}{\sqrt{2}}\right)$	$L(x')$	X	X	$y = xL(x')$	X

Fig. 2. Decomposition of non-linear functions at the level of operations. Each non-linear function is divided into three stages (except for GeLU, which has two stages).

Figure 2 illustrates the resource sharing of three non-linear functions in the ViT. After implementing the three-stage pipelined architecture, each stage (except stage 1) is modularized into basic computational units that can be shared among the different functions. In the modularized Non-Linear Unit, it is evident that basic operations such as subtraction, addition, multiplication, and accumulation (highlighted in yellow) can be shared. In terms of certain special computational units (highlighted in green), such as the maximum seeker and accumulator, exponential function and distribution function $L(x)$, and reciprocal function and square root reciprocal function, are designed properly to take advantage of the reuse concept. These units are designed with careful consideration to maximize their efficiency and effectiveness through following methods.

Adder and Comparator. In the context of signed computations, adders and subtractors can be considered unique units because subtracting a positive number is equivalent to adding its negative counterpart. On the other hand, comparing two numbers can be implemented by comparing each digit of the numbers or by using a subtractor followed by a multiplexer. Figure 3(a) shows a simple greater comparator using an adder. This comparator is a fundamental component of a maximum seeker and an accumulator. Figure 3(b) illustrates the implementation of the comparator with 8-bit inputs. The most significant bit (MSB) of adder's output is used as the input to a multiplexer, which selects the greater input. The reuse implementation shown in Fig. 3(c), which requires only two multiplexers and one adder, is advantageous for resource saving. This implementation efficiently switches between a normal adder and a greater seeker.

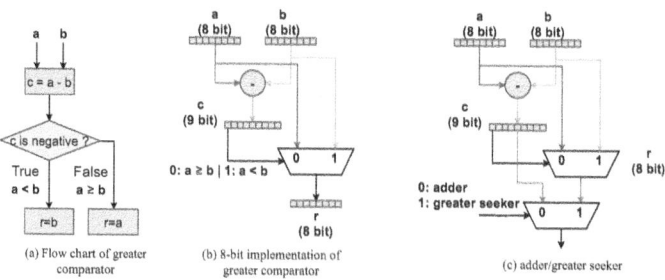

Fig. 3. Reuse technique for adder and greater seeker

Exponential Operation and Distribution Operation. To ensure that the pair of non-linear operations 2^x and $L(x)$ is implemented using identical hardware resources as mentioned in Fig. 2, these operations are approximated by second-order polynomials. Intuitively, by adjusting the parameters a, b, and c, the second-order polynomial unit can switch between multiple non-linear functions.

For the $L(x)$ function, the interpolation result from [2] demonstrated high accuracy in the original paper and was further validated in the recent study [3]. After verifying experimentally with our data, the formula for $L(x)$ is shown as follows:

$$\frac{1}{2}\left(1 + \frac{2}{\sqrt{\pi}} \int_0^x e^{-t^2} dt\right) = \begin{cases} 1 & x > 1.769 \\ -0.1444(1.769 - x)^2 + 1 & 0 < x \leq 1.769 \\ 0.1444(1.769 + x)^2 & -1.769 < x \leq 0 \\ 0 & x < -1.769 \end{cases} \quad (5)$$

In terms of the 2^x function, it is important to note that the expression $x'_j = \log_2 e(x_j - x_{max})$ in Eq. (1) is a non-positive real value. In case that real values in the design are represented in a fixed-point format $[n, m]$, where n and m are respectively the bit numbers for the integer and fractional parts, real values can be easily separated into these two parts. This allows the power of two of an integer to be considered as a shift operation, as shown:

$$2^{x'_j} = 2^{-(d_j+r_j)} = \left(2^{-r_j}\right) \gg d_j \quad \text{for } 0 \leq r_j < 1 \tag{6}$$

where d_j and r_j are the non-negative integer part and non-negative fractional part, respectively.

$$2^{x'_j} = \left(0.1714 \cdot (1.9472 - r_j)^2 + 0.3484\right) \gg d_j \tag{7}$$

The range of 2^{-r_j} is within $(0.5, 1]$, which is a small range compared to the domain of real numbers. Hence, it is suitable to apply a second-order polynomial approximation to this range, which helps to retain maximum accuracy. From this, the final equation for the exponential operation is shown in Eq. (7).

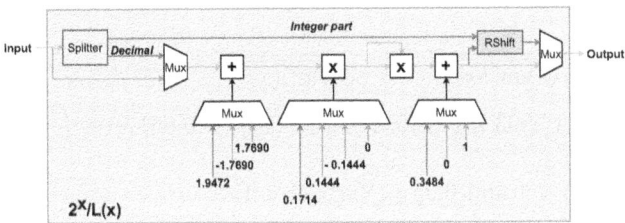

Fig. 4. The full design of the $2^x/L(x)$ unit, where the red bus, orange bus, and yellow bus represent the 2^x path, the $L(x)$ path, and the common path of these two operations, respectively. RShift denotes Right Shifter and Splitter is used to split the input into decimal and fractional parts. (Color figure online)

Reciprocal Operations. Retaining the idea from exponential operation approximation, before the approximation of the reciprocal operation $\frac{1}{s(x)}$ and the square root reciprocal function $\frac{1}{\sqrt{\text{Var}(x)}}$ in Fig. 2, the input range should be reduced to a smaller range. The reciprocal function for Softmax becomes:

$$\frac{1}{s(x)} = \frac{1 \times 2^{-16}}{s(x) \times 2^{-16}} = \frac{2^{-16}}{s'(x)} = \frac{1}{s'(x)} \gg 16 \quad \text{where } 0 \leq s'(x) < 1 \tag{8}$$

The real number $s(x)$ is represented in a 16-bit fixed-point format, with the maximum value being $2^{16} - 1$ (in the case of an unsigned fixed-point format $[n, m]$, where n and m are flexible based on the oriented design). Therefore, if this value is divided by 2^{16}, it certainly becomes a value in the range from 0 to 1. In fact, $s(x)$ does not necessarily need to be truely divided by 2^{16}; instead, $s(x)$ can be interpreted as $s'(x)$ without any adjustment or additional resource usage by simply considering $s(x)$ in format $[n, m]$ as $s'(x)$ in format $[0, 16]$. After that, at the end of the calculation, a constant 16-bit shift is conducted to correct the output. This constant shifting does not require any additional resources as well.

Furthermore, to scale down the range of the denominator even more, a Leading Zero Counter (LZC) is used. Specifically, $s'(x)$ is detected the number of leading zeros before the first meaningful bit, and $s'(x)$ is left-shifted until this meaningful bit becomes the MSB. As a result, $s'(x)$ becomes a number in the smaller range from 0.5 to 1. Thus, Eq. (8) is transformed as follows:

$$\frac{1}{s'(x)} \gg 16 = \frac{1 \times 2^a}{s'(x) \times 2^a} \gg 16 = \frac{1}{s''(x)} \ll (a - 16) \qquad (9)$$

$$= 2.6276 \times ((1.1136 - s''(x))^2 + 0.3746) \ll (a - 16) \qquad (10)$$

where $0.5 \leq s''(x) < 1$, and a is the number detected by the LZC.

Similar to the reciprocal function, the square root reciprocal function is also shifted into a range from 0.25 to 1 before the approximation. Hence, it becomes:

$$\frac{1}{\sqrt{\text{Var}(x)}} = \frac{1}{\sqrt{\text{Var}'(x)}} \gg 8 = \frac{1}{\sqrt{\text{Var}''(x)}} \ll \left(\frac{a}{2} - 8\right) \qquad (11)$$

$$= 1.5903 \times ((0.9991 - \text{Var}''(x))^2 + 0.6517) \ll \left(\frac{a}{2} - 8\right) \qquad (12)$$

where $0 \leq \text{Var}'(x) < 1$ and $0.25 \leq \text{Var}''(x) < 1$.

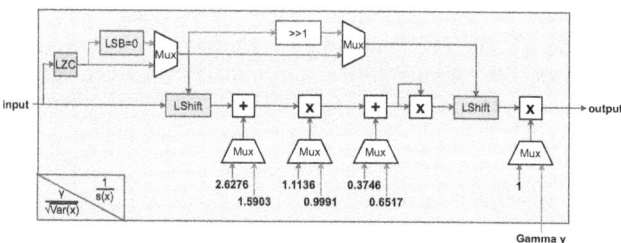

Fig. 5. The full design for reciprocal function and square root reciprocal unit, where the red bus, pink bus, and blue bus correspond to the Softmax path, the LayerNorm path, and the common path of these two functions, respectively. LShift denotes Left Shifter and $\gg 1$ is 1-bit Right Shifter. Noticeably, in the square root reciprocal mode, an additional simple unit is included to round the output of the LZC a to the even number by setting the LSB to 0. (Color figure online)

Entering into the details of each stage in processing architecture, in Fig. 6, apart from the basic operators (adder, multiplier, etc.), and special operators that can perform two types of operations, there are other components such as FIFO buffers and multiplexers (mux). Due to the reuse of most components in the architecture, multiplexers direct data to the appropriate components when they are operating in different functional modes. FIFO buffers retain one row of dependent data within a stage. Once the processing in a stage is complete (whether it's calculating the average, maximum, or an accumulated value), the buffers transfer the data row to the subsequent stage for further processing.

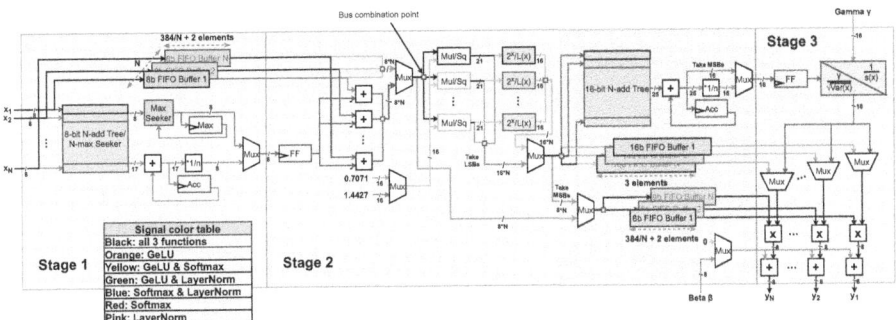

Fig. 6. The full architecture with colorized signal paths which indicate the dataflow of functions. Bus combination point indicates the position where bus of all threads is unified in illustration to simplify the diagram because they have the same dataflow.

4 Experimental Results

The proposed accelerator has been implemented using Vivado 2021.2 on an XCVU9P FPGA featuring 1,182K LUTs, 2,364K registers, and 6,8K DSP slices. The parallelism (processed elements per cycle) is 32, and the ViT-small model [3] is employed to accommodate the resource-limited device. The input and output data are represented as 8-bit integers, but they are converted to 16-bit fixed-point numbers for processing in real format. The 16-bit fixed-point format used is [5,11], selected based on the reference paper [7] and our experimental simulations. In general, the bit number for the integer part should cover the entire data range, while the remaining bits are reserved for the fractional part. The design consumes 3.063W, with a dynamic power consumption of 0.586W.

4.1 Accuracy

The metric is the L1 Norm, highlighting the difference between real values $y_i^{\text{Ground Truth}}$ and values launched from the proposed accelerator y_i^{Approx} at the bit level. Regarding the ground truth, experimental inferences are carried out to generate data. Samples are taken from different layers and inferences to ensure that the entire data range is covered. The formula for the L1 Norm is:

$$e_i = \left| y_i^{\text{Approx}} - y_i^{\text{Ground Truth}} \right| \quad (13)$$

where e_i represents the error of the i-th element.

Under the test conditions, both GELU and Softmax exhibit the majority of samples without errors, and the maximal error being 2 bits in the Least Significant Bits (LSBs). However, LayerNorm shows a greater accuracy loss, with a maximal error of 4 bits. In fact, although LayerNorm involves only a single approximation, $\frac{\gamma}{\sqrt{\text{Var}(x)}}$, this non-linear operation consists of two components: $\frac{1}{x}$ and $\sqrt{x}$, both of which are non-linear. This increases the complexity of the approximation, resulting in higher bit errors compared to Softmax and GELU.

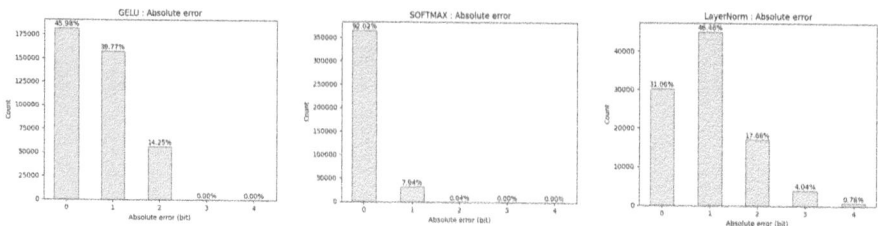

Fig. 7. The error histograms of the functions with the x-axis representing the error at the bit level and the y-axis counting the samples. The left figure shows the error histogram for GELU, the middle figure for Softmax, and the right figure for LayerNorm.

4.2 FPGA Synthesis

Table 1. Comparison of proposed architecture with state-of-the-art implementations. All works are implemented using 32 parallel threads and 16-bit data format.

Criteria		Ours	[3]	[4]
Function	Softmax	Yes	Yes	Yes
	LayerNorm	Yes	Yes	No
	GELU	Yes	Yes	Yes
Flexibility		Yes	No	Yes
Parallel Level		32	32	32
Resource	LUT	**7571**	43586	43940
	Register	**3657**	31800	44861
	DSP	134	234	0
Frequency		200 MHz	**300 MHz**	200 MHz
Time Processing (Layer Latency)	Softmax	**69.52 µs**	120 µs	-
	LayerNorm	**15.58 µs**	25 µs	-
	GELU	61.73 µs	-	-

With a parallelism level of 32 (32 elements processed per cycle), our design achieves a frequency of 200 MHz while utilizing only about 1% of the resources on the XCVU9P FPGA. Table 1 compares our accelerator with other state-of-the-art FPGA implementations. Compared to [4], although it eliminates the need for DSPs, arguing that DSPs might reduce the circuit speed, our design achieves the same operating frequency of 200 MHz as [4].

In terms of resource use, our design outperforms others for LUT and register utilization. Additionally, it is more optimized for resource balance, and it effectively supports three functions in ViT, compared to only two in [4]. Regarding the latency, our three-stage pipeline surpasses [3] in both Softmax and LayerNorm processing time, despite frequency differences. This indicates that our proposed design could replace existing non-linear functions implementations in other ViT accelerators, potentially improving the performance of all ViT inference.

5 Conclusion

The goal of this research is to propose a computational architecture for non-linear functions that can be embedded into any ViT model, inheriting all of its original non-linear functions. This architecture has been designed as a separate module, making it easily integrable with any matrix multiplication accelerator to form a complete ViT accelerator. In real-world applications, while accuracy loss may occur in the non-linear functions, the final inference accuracy of the entire model is the key consideration. Therefore, the entire model should be retrained with the approximated non-linear functions, updating parameters to minimize inference errors if applied to real-world scenarios in the future.

References

1. Vaswani, A., et al.: Attention is all you need. In: 31st Conference on Neural Information Processing Systems (NIPS 2017), Long Beach, CA, USA (2017). https://doi.org/10.48550/arXiv.1706.03762
2. Kim, S., et al.: I-BERT: integer-only BERT quantization (2023). https://doi.org/10.48550/arXiv.2101.01321
3. Huang, M., et al.: An integer-only and group-vector systolic accelerator for efficiently mapping vision transformer on edge. IEEE Trans. Circuits Syst. I: Regular Papers 70.12 (2023). https://doi.org/10.1109/TCSI.2023.3312775
4. Li, T., et al.: A high speed reconfigurable architecture for Softmax and Gelu in vision transformer. Electron. Lett. **59** (2023). https://doi.org/10.1049/ell2.12751
5. Ham, T.J., et al.: ELSA: hardware-software co-design for efficient, lightweight self-attention mechanism in neural networks. In: 2021 ACM/IEEE 48th Annual International Symposium on Computer Architecture (ISCA), Valencia, Spain (2021). https://doi.org/10.1109/ISCA52012.2021.00060
6. Khan H., et al.: NPE: an FPGA-based overlay processor for natural language processing. In: The 2021 ACM/SIGDA International Symposium on Field-Programmable Gate Arrays (2021). https://doi.org/10.1145/3431920.3439477
7. Koca, N. A., Do, A. T., Chang, C.-H.: Hardware-efficient Softmax approximation for self-attention networks. In: 2023 IEEE International Symposium on Circuits and Systems (ISCAS), Monterey, CA, USA (2023). https://doi.org/10.1109/ISCAS46773.2023.10181465
8. Chen, C., Li, L., Sabry Aly, M. M.: ViTA: a highly efficient dataflow and architecture for vision transformers. In: 2024 Design, Automation & Test in Europe Conference & Exhibition (DATE), Valencia, Spain (2024). https://doi.org/10.23919/DATE58400.2024.10546565
9. Yu, J., et al.: NN-LUT: neural approximation of non-linear operations for efficient transformer inference (2024). https://doi.org/10.48550/arXiv.2112.02191

LiFT: Lightweight, FPGA-Tailored 3D Object Detection Based on LiDAR Data

Konrad Lis[✉][iD], Tomasz Kryjak[iD], and Marek Gorgoń[iD]

Embedded Vision Systems Group, Department of Automatic Control and Robotics, AGH University of Krakow, Al. Mickiewicza 30, 30-059 Krakow, Poland
{kolis,tomasz.kryjak,mago}@agh.edu.pl

Abstract. This paper presents LiFT, a lightweight, fully quantized 3D object detection algorithm for LiDAR data, optimized for real-time inference on FPGA platforms. Through an in-depth analysis of FPGA-specific limitations, we identify a set of FPGA-induced constraints that shape the algorithm's design. These include a computational complexity limit of 30 GMACs (billion multiply-accumulate operations), INT8 quantization for weights and activations, 2D cell-based processing instead of 3D voxels, and minimal use of skip connections. To meet these constraints while maximizing performance, LiFT combines novel mechanisms with state-of-the-art techniques such as reparameterizable convolutions and fully sparse architecture. Key innovations include the Dual-bound Pillar Feature Net, which boosts performance without increasing complexity, and an efficient scheme for INT8 quantization of input features. With a computational cost of just 20.73 GMACs, LiFT stands out as one of the few algorithms targeting minimal-complexity 3D object detection. Among comparable methods, LiFT ranks first, achieving an mAP of 51.84% and an NDS of 61.01% on the challenging NuScenes validation dataset. The code will be available at https://github.com/vision-agh/lift.

Keywords: 3d object detection · FPGA · LiDAR · NuScenes · quantisation

1 Introduction

With the rapid development of autonomous vehicle technology, one of the key challenges becomes ensuring reliable perception of the surroundings. In particular, real-time, high-precision 3D objects detection is the foundation for the safety and efficiency of autonomous systems. Among the various sensors used in autonomous vehicles, LiDAR (Light Detection and Ranging) has gained considerable popularity due to its ability to generate accurate 3D maps of the environment, allowing for precise object detection and classification. Thanks to its independence from lighting conditions and its high resolution and accuracy, the technology offers advantages over other sensors such as cameras and radars.

Modern 3D object detection algorithms based on LiDAR data typically use Deep Convolutional Neural Networks (DCNNs), which offer high performance but come with significant computational and memory demands. As a result, they are usually implemented on high-performance computers with GPU (Graphics Processing Unit) cards for efficient training and inference. However, the overarching goal of these algorithms is real-time operation in autonomous vehicles or advanced driver assistance systems (ADAS). For application in an actual, mass-produced vehicle, a reliable, power-efficient and low-cost computing platform is required – for example a modern SoC FPGA. The real-time reimplementation of SoTA (State of The Art) algorithms on such platforms is a major challenge and often requires a significant redesign of the algorithm to take full advantage of the capabilities of the platform under consideration (so-called hardware aware algorithm design).

In this paper, we present LiFT, a lightweight 3D detector based on LiDAR sensor data, carefully designed to run in real-time on low-power FPGA or ASIC platforms while providing high detection performance on the demanding NuScenes dataset by combining a number of novel solutions with SoTA mechanisms.

The main contributions of this paper are:

- a set of constraints on a 3D detector architecture in the context of implementation on an FPGA platform,
- an efficient way to quantize initial features,
- Dual-Bound Pillar Feature Net – a Pillar Feature Net extension to increase detection performance without added complexity,
- a 3D detection algorithm adapted for implementation in FPGAs, with the best detection efficiency among comparable methods on the NuScenes dataset.

The reminder of this paper is organised as follows. In Sect. 2 we discuss issues related to our work: most commonly used datasets, DCNN approaches to object detection in LiDAR data and FPGA/ASIC implementations of such algorithms. Next, in Sect. 3 we elaborate on the LiFT design, focusing on hardware induced constraints on the algorithm and novel mechanisms. The results obtained are summarised in Sect. 4. The paper ends with a short summary with conclusions and discussion of possible future work.

2 Related Work

Datasets. The most commonly used datasets for object detection based on LiDAR data are KITTI [6] (2012), NuScenes [3] (2019) and Waymo [14] (2019). The latter two sets are much larger and more challenging than KITTI. NuScenes contains 1,000 sequences that add up to 1.4 million images, 390,000 LiDAR scans and 1.4 million tagged objects. Of the 390k LiDAR scans, only 40k are labelled – 28310 are used for training, 6019 for validation, and 6008 for testing. 3D object detection on the NuScenes is evaluated using the standard mAP metric (mean Average Precision) and a metric called NDS (nuScenes detection score). It includes mAP and several error measures, e.g. orientation error or scale error.

Methods. There are several approaches to 3D object detection from LiDAR data, the most popular of which are point-based and cell-based methods (the later most common in SoTA solutions). In the first step, the point cloud is reduced to a regular grid of 2D or 3D cells (pillars or voxels). The cells are assigned a feature vector using the Pillar Feature Encoder (PFE) in the 2D case or the Voxel Feature Encoder (VFE) in the 3D case. Due to the rationale described in Sect. 3.1, the review is limited to detectors using 2D cells.

The first and the simplest solution is PointPillars [7], with the PFE called Pillar Feature Net (PFN), consisting of a linear layer, Batch Normalization and ReLU, which are applied to each point individually. Points from each pillar are reduced to a feature vector using pointwise *Max pooling*. Another 2D detector is CenterPoint-Pillar [15], which uses PointPillars as a backbone and a novel CenterHead. In the one-stage version, it predicts a heatmap with the centres of objects and regression maps specifying their location, shape and orientation. Both PointPillars and CenterPoint-Pillar are based on regular convolutional neural networks. However, usually 90% of the input pillars are empty and are implicitly assigned a zeroed feature vector, what can be exploited using sparse convolutional neural networks to significantly reduce computations. In addition to standard sparse layers, which function similarly to regular convolutions by gradually expanding the area of non-zero, so-called active, pixels, there are also submanifold versions of these layers, which maintain the active pixels structure intact. Other popular 2D detectors, representing the SoTA on datasets such as NuScenes and Waymo, include PillarNet [12] and its derivatives VoxelNeXt-2D [4] and FastPillars [16]. VoxelNeXt-2D is particularly notable as it incorporates only sparse convolutions, even in its head.

FPGA/ASIC-Based Implementations. To date, there have been few implementations of LiDAR sensor-based object detectors on FPGA platforms, all of which are based on the original or modified version of PointPillars [1,2,8,10,13]. Most of them operate in real-time, defined as processing 10 point clouds per second, which corresponds to the typical LiDAR rotation frequency. All of them are evaluated on the KITTI dataset. However, methods capable of performing well on more challenging datasets, such as the NuScenes are currently missing.

The authors of SPADE [9] and SPADE+ [11] undertook the task of developing a 2D sparse convolution accelerator targeting an ASIC, which was evaluated in simulation. In addition, they implemented PointPillars, CenterPoint-Pillar and PillarNet in several different sparse versions and tested the accelerator performance. PointPillars was evaluated on KITTI, while CenterPoint-Pillar and PillarNet were evaluated on NuScenes.

3 Hardware-Aware Design of LiFT

3.1 Hardware-Induced Constraints on Algorithm Design

Our goal is to develop a 3D detection algorithm with relatively high detection accuracy, suitable for real-time implementation on an FPGA platform. As a reference platform, we will use the AMD/Xilinx Kria K26 SOM platform, a mid-range SoC FPGA device, which could serve as an embedded processing platform for LiDAR data. In the design process, it is essential to consider the constraints imposed by the chosen computational platform, particularly in terms of computational complexity, memory usage, and other algorithmic factors. It is important to emphasize that the algorithm will not be restricted to this specific platform and could be adapted for use on other FPGA or ASIC devices as well.

The first issue to address, when considering computational complexity, is the precision of the calculations. Floating-point operations are more resource-intensive and time-consuming, so, whenever possible and without significant loss of accuracy, it is recommended to perform quantization of the algorithm's computations. The most common approach is 8-bit integer quantization (INT8), which, when combined with quantization aware training, provides sufficient detection accuracy with high processing speed and low memory consumption. In addition, INT8 quantization is widely supported by most FPGA accelerators and other embedded systems. For these reasons, this work opts to use INT8 quantization.

To determine the upper limit on the computational complexity of the 3D detection algorithm, we refer to the DNN accelerator provided by AMD/Xilinx - the DPU (Deep Learning Processor Unit). The computational complexity of the neural network will be expressed in terms of MAC (Multiply And Accumulate) operations, representing the number of multiply-accumulate operations required to produce the network's output. The K26 SOM platform's logic resources can accommodate a single instance of the B4096 version of the DPU, operating at a clock frequency of 300 MHz. This means that the DPU can perform 2048 MAC operations per clock cycle, resulting in a processing rate of 614.4 GMAC/s ($1 GMAC = 10^9 MAC$) at a clock speed of 300 MHz. Assuming a typical real-time processing definition, i.e., 10 point clouds per second (pcd/s), the number of operations required for a single point cloud, in the ideal case, must not exceed 61.44 GMAC. In practice, however, determining the exact upper limit for number of operations is not trivial, as all the algorithms include operations that are not included in the assessment of computational complexity. According to [1], among the algorithms capable of processing 10 frames per second, the maximum observed complexity is 50 GMAC, an average is around 30 GMAC. Adopting a limit of 50 GMAC could be overly optimistic, given the potential overheads of LiFT such as dividing point cloud into 2D cells, therefore we assume an upper bound on computational complexity of 30 GMAC. Of course, the lower the computational complexity of the algorithm, the better. However, reducing complexity is not always justified if it leads to a significant drop in detection performance. Ultimately, the algorithm's runtime should be verified on the target platform. The adopted upper limit of 30 GMAC is merely an approximation, indicating which 3D detection algorithms, operating on LiDAR data, are worth implementing in an embedded system if real-time processing is to be achieved.

When addressing the issue of memory complexity, it is important to first consider the specifics of FPGA systems. For AMD/Xilinx FPGAs from the Zynq Ultrascale+ series, the system includes both programmable logic with a small amount of dedicated memory – referred to as on-chip memory (OCM), typically up to a few MiB (SRAM), and a processing system with external DRAM memory of several GiB. In data processing within programmable logic, it is preferable to use on-chip memory due to its short and predictable access time, which amounts to only a few clock cycles. In contrast, operations involving DRAM take a non-deterministic number of cycles – tens or more, and they consume significantly more energy compared to operations performed on OCM.

In the context of designing 3D object detectors, the limited amount of on-chip memory becomes a critical factor influencing the structure of the entire algorithm. Specifically, two main aspects of the detector design are particularly sensitive to this limitation. The first is the amount of skip connections, which require storing entire data tensors. Due to the small capacity of the OCM, tensor often needs to be stored in the external DRAM. This results in increased detector latency because of additional data transfers, especially when the number of skip connections is large. Thus, it is recommended to use architectures with no skip connections or only a very limited number of them.

The second aspect is the dimensionality of processed cells. DNN accelerators on FPGA, for each convolutional layer, must perform the *Im2Col* operation, which generates the convolutional context. It requires buffering the pixels that constitute subsequent contexts, and for efficiency, OCM is preferred for this task. In the case of a dense 3D tensor with dimensions XY of 640×720 and a Z-height of 40, assuming a context size of $3 \times 3 \times 3$, a buffer of $40 \times 640 \times 2 + 640 \times 2 + 3 = 52483$ cells is required. In comparison, for a 2D tensor with the same XY dimensions and a context size of 3×3, only $640 \times 2 + 3 = 1283$ cells are needed – over 40x fewer than in the 3D case. Given the limited on-chip memory, it is therefore advisable to use detectors operating on 2D cells rather than 3D cells.

In summary, the computational platform and the requirement for real-time processing impose several constraints on the detection algorithm. The limited amount of on-chip memory necessitates operating on a two-dimensional grid of cells and restricting the number of skip connections to a maximum of a few. Meanwhile, considerations related to the computational power and logical resources of the FPGA mandate the use of INT8 quantization and limit the computational complexity to 30 GMAC.

3.2 Constraints Analysis and Design Assumptions

When looking for an architecture that satisfies the constraints described in the Sect. 3.1, the detection efficiency should be maximised at the same time. All of the aforementioned limitations, in the general case, result in a reduction of the precision and amount of information available in the detector or potentially a less efficient learning process. In this subsection we analyse adopted constraints and establish additional assumptions to achieve a high performance detector.

Skip connections are widely used in neural network architectures to address the vanishing gradient problem during training and to combine features across different scales. Recently, the RepVGG approach [5] has been introduced, allowing the use of skip connections spanning across single layers during training while removing them during inference through a simple weight transformation. In designing our computational architecture, we will leverage the RepVGG mechanism to eliminate most skip connections. However, we will retain residual connections related to multi-scale feature fusion, as their removal would result in a significant drop in detection accuracy, as demonstrated in Sect. 4.2.

When processing data as 2D cells instead of 3D cells, constructing an effective feature vector for each cell becomes more challenging, as all dimensions of input data must be preserved due to detection in three dimensions. For 3D cells a common approach is to average point features in a given cell, however it would result in a significant height information loss in the 2D cells case. For this reason, more advanced solutions are used. The most common encoder is PFN from [7], which is used in many other detectors such as [4,12], or its extensions, such as MAPE from [16]. In our architecture, we will use a custom extension of PFN – DBPFN – described in more detail in Sect. 3.3. By employing *Min pooling* alongside *Max pooling*, our solution minimally complicates the PFN computational architecture while significantly increasing its effectiveness.

Another important aspect to analyse is computation quantization. We intend to use INT8 quantization, so it is crucial that the input features are well represented in 8 bits to avoid losing too much information at the beginning. Thanks to the method described in Sect. 3.4, we leverage both the convenience of using 8-bit quantisation everywhere in LiFT and the effective localization resolution below 2 mm compared to 40 cm in a default quantisation case.

Considering the computational complexity and referring to contemporary 3D detectors, we plan to use sparse convolutions. These reduce computational complexity by an average of 50% to 80%, depending on the specific architecture. However, achieving this reduction entails an additional cost due to the more complex context generation process.

In conclusion, having analysed the constraints from Sect. 3.1, we propose using short skip connections as introduced in RepVGG, encoding features for pillars with DBPFN, implementing efficient input feature quantization, and utilizing sparse convolutions.

3.3 Dual-Bound Pillar Feature Net

The structure of the original PFN is simple and relatively efficient. It is commonly used, e.g. in PillarNet [12] and VoxelNeXt [4]. There is potential to increase its efficiency through various extensions, but this comes with the added complexity of the architecture, for instance, by introducing additional operation like self-attention as seen in MAPE [16]. In our work, we propose an extended version of PFN, designed in such a way as to preserve its simplicity. This allows us to avoid complicating the FPGA implementation by introducing new operators.

The PFN, despite its relative efficiency and popularity, has several downsides. One is the significant reduction in information about the distribution of points in the pillar. Features from all points are reduced to a single vector by the *Max pooling* operation, thus potentially losing most of the information about

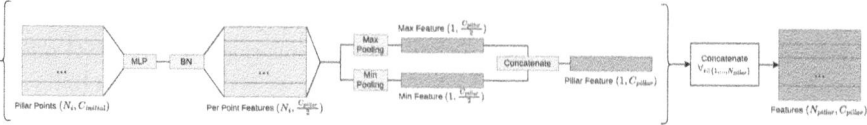

Fig. 1. An outlook on Dual-Bound Pillar Feature Net (DBPFN) structure

the distribution of features within the pillar. Another issue is the blockage of gradient flow to the MLP due to the *ReLU* activation function and *Max Pooling*. Through the *Max Pooling* operation, each output feature from the MLP is only learned based on one point within the given pillar – specifically, the one that happened to have the largest value of the feature. Additionally, if the data and weight distributions cause the activation entering *ReLU* to be negative in most cases, the neuron will fall into the region of *ReLU* with a zero slope. This results in a zero gradient for that feature in the MLP, leading to what is known as a "dead neuron" which will stop getting updates during training.

These issues can be partially addressed by a small modification to PFN that only slightly complicates the architecture – removing the *ReLU* activation function, using *Min Pooling* alongside *Max Pooling*, and concatenating the features obtained from both operations. We call this modified PFN the Dual-Bound Pillar Feature Net (DBPFN), and its structure is shown in Fig. 1. By incorporating *Min Pooling* in addition to *Max Pooling*, the individual MLP features are learned from two points within the pillar instead of just one, preserving more information about the feature distribution inside the pillar. This is not a major quantitative change, but selecting the minimum point alongside the maximum potentially diversifies the set of points on which each MLP feature is learned. With the addition of *Min Pooling*, removing *ReLU* becomes essential, as *ReLU* would zero out negative values, causing *Min Pooling* to return zeros more often instead of more descriptive feature values. To maintain the same number of output features in the DBPFN as in the original implementation, the number of features in the MLP must be reduced by a half. However, as it will be shown in Sect. 4.2, our implementation is still more efficient than the original PFN.

3.4 Effective Input Features Quantisation

We opt to apply INT8 quantization, so in order to minimize information loss, the input features must be well represented with 8 bits. Detectors working on the NuScenes dataset typically use a range of -54 m to 54 m in both the X and Y dimensions. Assuming we want to utilize the full range, we would achieve a resolution of about 40 cm with 8-bit quantization. This is far too coarse compared to the size of the pillars, which is 15 cm x 15 cm in our case. For this reason, we divided each of the XYZ dimensions into two features: a coarse and a detailed location. For the X dimension, the definitions of the features X_{coarse} and X_{detail} are as follows:

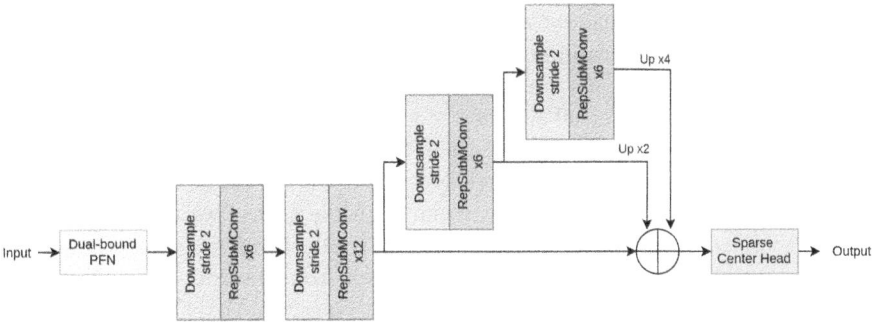

Fig. 2. An outlook on LiFT structure

$$X_{coarse} = \left\lfloor \frac{X}{resolution_X} \right\rfloor \times resolution_X \qquad X_{detail} = X - X_{coarse}$$

where $resolution_X = 2^{-8} * (X_{max} - X_{min})$, X denotes the location of a given point along the X-axis, and X_{min} and X_{max} define the boundaries of the point cloud along the same axis. In a similar manner, the features Y_{coarse}, Y_{detail}, Z_{coarse}, and Z_{detail} are defined.

3.5 LiFT Design

The schematic of our proposed architecture – LiFT – is shown in Fig. 2. The detector operates on 2D cells and consists entirely of sparse convolutions, including the head. All weights and activations are quantized with type INT8. We have also used the effective quantization of input features described in Sect. 3.4. As the PFE, we use DBPFN, described in Sect. 3.3.

To remove short skip connections during inference, we use a reparameterisation scheme from RepVGG. Thus, we define an ordinary reparametrisable sparse

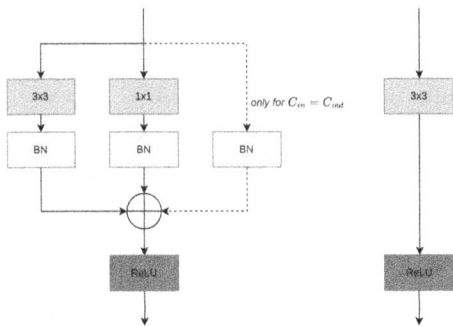

Fig. 3. Reparametrisable convolution structure during training (on the left) and during inference (on the right).

convolution layer – *RepSparseConv* and its submanifold version *RepSubMConv*. Their structure during training and inference is shown in Fig. 3. In the backbone, we use 4 stages – each stage consisting of a downsampling layer based on RepSparseConv with a kernel size of 3 × 3 and stride equal 2, and a number of RepSubMConv layers with a kernel size of 3 × 3 and stride equal 1. The number of RepSubMConv layers in each stage is 6, 12, 6, and 6, respectively.

When it comes to fusing features from multiple scales, we use an approach inspired by the one applied in VoxelNeXt 2D [4]. This involves adding the output tensors from the 3rd and 4th stages to the tensor from the 2nd stage, previously upsampling them to the resolution of the 2nd stage. In VoxelNeXt 2D, tensor upsampling from the 3rd and 4th stages consisted of simply multiplying the coordinates of the sparse tensor by 2 and 4, respectively. However, this procedure created only one pixel while the resolution changed 2x or 4x. Moreover, in general case, it did not align with the corresponding active pixels in the second stage. Therefore, we introduced a different strategy for changing the resolution. When adding tensors, we keep the active pixels from the second stage only and add the corresponding pixels from the third and fourth stages to them. Our solution is more efficient than the VoxelNeXt's one, as it will be shown in Sect. 4.2. The head design, on the other hand, is borrowed from the VoxelNeXt – it is a Sparse Center Head based entirely on sparse convolutions.

4 Experiments

The network was trained on the NuScenes dataset. We use a pillar size of 15cm x 15cm and a point cloud range of [−54 m, 54 m] along the X and Y axes and [−5 m, 3 m] along the Z axis. The number of features in the PFE output is 64, while in each respective stage is 64, 64, 128, 128. In the output of the second stage, we apply an additional layer to align the number of features with the 3rd an the 4th stage. When it comes to data augmentation and training hyperparameters, we follow the VoxelNeXt settings. We ran the training on a computer with 7 Nvidia RTX A6000 cards. The total batch size was equal to 42.

4.1 Overall Results

LiFT evaluation results are presented in Table 1. It was compared with other detectors evaluated on the NuScenes, which meet the constraints defined in the Sect. 3.1. The only detectors of this kind, except for LiFT, are different sparse versions of the CenterPoint, presented in [9,11]. To the best of our knowledge, no other work has presented a detector that meets the aforementioned constraints. Nor are we aware of any other detector with a complexity of less than 30 GMAC, even after relaxing the other limitations. LiFT achieved 51.84% mAP and 61.01% NDS on the NuScenes-val set with a computational complexity of 20.73 GMAC. In terms of mAP and NDS metrics, it ranks first, with a large margin of 9.27 GMAC from the computational complexity threshold.

Table 1. Performance comparison of 3D object detection algorithms based on 2D cells with number of operations equal at most 30 GMAC.

Source	Detector	mAP [%]	NDS [%]	GMAC
ours	**LiFT**	**51.84**	**61.01**	20.73
SPADE [9]	SCP2	50.12	60.42	24.77
	SCP3	47.78	58.97	**13.60**
SPADE+ [11]	SparseCenterPoint - SubM-Conv	47.89	58.94	18.13
	SparseCenterPoint - PS-Conv	50.12	60.42	27.09
	SparseCenterPoint - FS-Conv	50.30	60.41	23.34
	SparseCenterPoint - SD-Conv	50.33	60.84	19.37

Relative to the second best solution, *SparseCenterPoint - SD-Conv* from [11], LiFT is better by 1.51% in terms of the mAP metric and 0.17% in terms of NDS, respectively. At the same time, LiFT has a slightly higher computational complexity – by 1.36 GMAC. As we mentioned in Sect. 3.1, even with real-time processing achieved, reducing computational complexity is still desirable. However it can not result in a significant drop of mAP and NDS metric. A potential 1.36 GMAC reduction in computational complexity, obtained by choosing the second best solution, could be justified in terms of the NDS metric - the decrease is only 0.17%. In contrast, the mAP metric shows a much more considerable decline of 1.51%. For this reason, we believe that according to the constraints and assumptions made, LiFT is the best choice for implementation on an FPGA platform.

To conclude the discussion and provide a comprehensive comparison with SoTA methods, we now focus exclusively on detection performance, disregarding all other factors. Although LiFT falls significantly short of SoTA detection accuracy under unconstrained conditions, it still outperforms a considerable 12% of all submissions (in January 2025) on the NuScenes [3] leaderboard, including PointPillars [7] with mAP of 30.5% and NDS equal to 45.3%.

4.2 Ablation Studies

In the ablation study, we analysed the impact of removing multi-scale processing as well as four distinct components of LiFT that are crucial to its high detection performance. The result of experiments, proving the effectiveness of the four incorporated solutions, are presented in Table 2. We compared the DBPFN against the implementation with the original PFN – of the components analysed, it has the greatest impact on detection efficiency. The effective quantization of the initial features was challenged against a naive solution in which none of the features are split and all are quantized with INT8 type. This element has a similar impact on detection performance as the introduction of reparametrisable convolutions. We compared their efficiency against architectures with the same number of layers, but based on a residual connection structure called BasicBlock,

Table 2. Effects of different components of LiFT on mAP and NDS

Efficient scale fusion	Reparametrisable convolutions	Efficient quantisation	Dual-bound PFN	mAP [%]	NDS [%]
✓	✓	✓	✓	**51.84**	**61.01**
✗	✓	✓	✓	51.49 (−0.35)	60.73 (−0.28)
✓	✗	✓	✓	51.05 (−0.79)	60.63 (−0.38)
✓	✓	✗	✓	51.13 (−0.71)	60.47 (−0.54)
✓	✓	✓	✗	50.29 (−1.55)	59.87 (−1.14)

commonly used in 3D detectors, such as [4,12]. It turns out that the introduction of reparametrizable convolutions instead of classical skip connection structures does not only allow to obtain a simple structure during inference, but also positively affects the detection efficiency. The last element examined is the effective scale fusion described in Sect. 3.5, which was compared against an implementation from VoxelNeXt 2D. Our solution has slightly better detection performance while as simple as the one from VoxelNeXt. In addition, it allowed us to make a small but noticeable decrease in computational complexity by reducing the number of pixels that enter the Sparse CenterHead.

We conducted an additional experiment in which we removed multiscale processing from LiFT. Detection efficiency, as measured in terms of both mAP and NDS, dropped significantly – by 9.43% and 6.69%, respectively. Potentially, removing the last two skip connections would allow a slight speed-up in network performance by removing a few transfers between the FPGA and the external memory and reducing external memory consumption. However, the gain is so small that it is not justified with such a large decrease in efficiency, so we decided to keep the multiscale processing.

5 Conclusion and Discussion

In this paper, we present **LiFT**, a 3D object detection algorithm tailored for real-time implementation on FPGA. LiFT has been carefully designed, starting with determining the hardware induced constraints. In the next step we introduced Dual-Bound Pillar Feature Net and an efficient scheme to quantize INT8 input features so as to provide high detection performance while fulfilling the limitations. By combining a holistic analysis of the problem with our novel mechanisms and SoTA solutions, such as reparametrizable convolutions and fully sparse architecture, we obtained an algorithm with the best detection performance among methods of comparable complexity. We achieved 51.84% mAP and 61.01% NDS on the NuScenes-val set with a computational complexity of 20.73 GMAC. We believe we have developed a solid baseline for implementation on FPGAs and we hope our work will encourage the research community to pursue the topic of designing 3D detectors for embedded devices more often.

In future work, we plan to implement LiFT on FPGA. In addition, we plan to further improve the LiFT architecture, e.g. by using Spatially-Dilated Sparse Convolutions from SPADE+, which, according to the authors, significantly increase detection efficiency due to an increased reception field with little computational complexity overhead.

Acknowledgements. The work presented in this paper was supported by the AGH University of Krakow, project no. 10.16.120.79990 and the program "Excellence initiative – research university" for the AGH University of Krakow.

Disclosure of Interests. The authors have no competing interests to declare that are relevant to the content of this article.

References

1. AMD/Xilinx: Vitis AI model Zoo. https://xilinx.github.io/Vitis-AI/3.0/html/docs/workflow-model-zoo.html. Accessed 22 Nov 2024
2. Brum, H., Véstias, M., Neto, H.: LiDAR 3D object detection in FPGA with low bitwidth quantization. In: Applied Reconfigurable Computing. Architectures, Tools, and Applications, pp. 90–105. Springer Nature Switzerland, Cham (2024)
3. Caesar, H., Bankiti, V., Lang, A.H., Vora, S., et al.: nuScenes: a multimodal dataset for autonomous driving. arXiv preprint arXiv:1903.11027 (2019)
4. Chen, Y., Liu, J., Zhang, X., Qi, X., et al.: VoxelNeXt: fully sparse VoxelNet for 3D object detection and tracking. In: Proceedings of the IEEE/CVF Conference on Computer Vision and Pattern Recognition (CVPR), pp. 21674–21683 (2023)
5. Ding, X., Zhang, X., Ma, N., Han, J., et al.: RepVGG: making VGG-style convnets great again. In: Proceedings of the IEEE/CVF Conference on Computer Vision and Pattern Recognition (CVPR), pp. 13733–13742 (2021)
6. Geiger, A., Lenz, P., Stiller, C., Urtasun, R.: Vision meets robotics: the KITTI dataset. Int. J. Robot. Res. (IJRR) (2013)
7. Lang, A.H., Vora, S., Caesar, H., et al.: PointPillars: fast encoders for object detection from point clouds. In: 2019 IEEE/CVF Conference on Computer Vision and Pattern Recognition (CVPR), pp. 12689–12697 (2019)
8. Latotzke, C., Kloeker, A., Schoening, S., Kemper, F., et al.: FPGA-based acceleration of lidar point cloud processing and detection on the edge. In: 2023 IEEE Intelligent Vehicles Symposium (IV), pp. 1–8 (2023)
9. Lee, M., Park, S., Kim, H., Yoon, M., et al.: SPADE: sparse pillar-based 3D object detection accelerator for autonomous driving. In: 2024 IEEE International Symposium on High-Performance Computer Architecture (HPCA), pp. 454–467 (2024)
10. Li, X., Ren, A., Tan, Y., Li, X., et al.: VEA: an FPGA-based voxel encoding accelerator for 3D object detection with lidar. In: 2022 IEEE 40th International Conference on Computer Design (ICCD), pp. 509–516 (2022)
11. Park, S., Lee, M., Choi, J., Choi, J.: Selectively dilated convolution for accuracy-preserving sparse pillar-based embedded 3D object detection. In: Proceedings of the IEEE/CVF Conference on Computer Vision and Pattern Recognition (CVPR) Workshops, pp. 8104–8113 (2024)
12. Shi, G., Li, R., Ma, C.: PillarNet: real-time and high-performance pillar-based 3D object detection. In: Computer Vision – ECCV 2022, pp. 35–52. Springer Nature Switzerland, Cham (2022)

13. Stanisz, J., Lis, K., Gorgon, M.: Implementation of the pointpillars network for 3D object detection in reprogrammable heterogeneous devices using FINN. J. Signal Process. Syst. (2021)
14. Sun, P., Kretzschmar, H., Dotiwalla, X., Chouard, A., et al.: Scalability in perception for autonomous driving: Waymo open dataset (2019)
15. Yin, T., Zhou, X., Krahenbuhl, P.: Center-based 3D object detection and tracking. In: Proceedings of the IEEE/CVF Conference on Computer Vision and Pattern Recognition (CVPR), pp. 11784–11793 (2021)
16. Zhou, S., Tian, Z., Chu, X., Zhang, X., et al.: FastPillars: a deployment-friendly pillar-based 3D detector (2023). https://arxiv.org/abs/2302.02367

Efficient Processing using AI for Image, Vision and Signal Applications

A Practical HW-Aware NAS Flow for AI Vision Applications on Embedded Heterogeneous SoCs

Agathe Archet[1,2,3]([✉]), Nicolas Ventroux[1], Nicolas Gac[2], and François Orieux[3]

[1] cortAIx Labs, Thales Research & Technology, Palaiseau, France
agathe.archet@thalesgroup.com
[2] Université Paris-Saclay, CNRS, ENS Paris-Saclay, Laboratoire SATIE, Gif-sur-Yvette, France
[3] Université Paris-Saclay, CNRS, CentraleSupélec, Laboratoire L2S, Gif-sur-Yvette, France

Abstract. Implementing efficient *Deep Neural Networks* (DNNs) for dense-prediction vision applications on embedded heterogeneous SoCs comes with many challenges, such as latency and energy constraints. To tackle them, we propose a novel and practical multi-objective *Hardware-aware Neural Architecture Search* (HW-NAS) framework able, for the first time, to handle complex search spaces while considering the hardware manufacturer's expertise. This HW-NAS flow targeting Nvidia's Orin SoCs relies on (1) a practical strategy to reduce the total exploration duration, and (2) a compact enhancement of the existing TensorRT deployment flow. On the FasterSeg's search space, our framework can obtain a latency-power-mIoU Pareto front for multiple power modes in only 66 h (-33 % than the inital flow) using 8 Nvidia A100 GPUs. Compared to default mappings, these results demonstrate that our novel mapping strategy can obtain practical solutions with either 50 % less power consumption or 80 % less latency for the same accuracy performance, or achieve a better accuracy (+6 %) with 30 % less power consumption.

Keywords: Neural Architecture Search · Heterogeneous computing · Neural Networks · Low-power inference · DLA · semantic segmentation · Jetson Orin

1 Introduction

Current AI-based solutions allow state-of-the-art performances for dense vision applications (e.g., segmention, detection) compared to traditional methods. For edge computers, best image processing *Deep Neural Networks* (DNNs) are based on sophisticated neural architectures to get efficient accurate models [13]. For instance, real-time lightweight semantic segmentation [11] DNNs rely on specialized architecture strategies to aggregate multi-resolution features, such as encoder-decoder or multi-branch architectures (BiseNetV2 [28], FasterSeg [8]).

Hence, these specialized DNNs come at the cost of an increased architectural *complexity*. This results in more varied execution scenarios, that the computation processors have to accommodate. Thus, the use of DNNs leads to a very high computing complexity and opens up many challenges for industrialists that needs to design energy-efficient solutions for the inference stage.

Firstly, the selected hardware computer has a direct impact on the final system frugality (i.e. used hardware resources and final performances). For instance, low-power accelerators such as *Neural Processing Units* (NPUs), like the DLA [18], can deliver the best energy efficiency thanks to their dedicated hardware but comes with a reduced programmability and increased development costs [14]. On the opposite, *Graphical Processing Units* (GPUs) [18] offer an excellent programmability but their energy efficiency is limited by their general-purpose architecture. Otherwise, heterogeneous *System-on-Chips* (SoCs), such as the *Nvidia Jetson Orin Series* [18], allow the developer to find the right compromise between programmability and energy efficiency but at the cost of heavier configurations and complex co-design choices (e.g. mapping) to optimally use available hardware resources, which favor longer deployment time.

Secondly, one must rely on industrial tools and libraries based on the manufacturer's expertise to reach the best performances. However, some of these tools combined with certain SoC targets are proprietary since they benefit from the manufacturer's expertise and act as black-box processes, e.g. *TensorRT* for Nvidia SoCs. Consequently, they can bring useful extra optimizations that cannot be manually reproduced or easily deducted. Also, their use has an impact on time predictability and system sizing since the system behavior cannot be defined before compilation. In particular, TensorRT closely decides how resources are allocated. Besides, to capture the true hardware behaviour during execution, real physical measurements are preferred to abstract estimation proxies.

Finally, *handcrafted* DNN architectures easily lead to sub-optimal solutions [17], due to their settings complexity notably in the case of dense prediction tasks, but also because of the hardware configurability. Tailoring a frugal AI application requires to acknowledge both the hardware and the neural networks' specific levers to maximize energy efficiency. With these two complex design spaces (HW and DNN configurations), only an automated joint optimized *Design Space Exploration* (DSE) can consider both levers. *Hardware-aware Neural Architecture Search* (HW-NAS) methods can leverage the available hardware resources and obtain the best compromise between application performance and energy efficiency [3]. Still, their efficiency is restricted by long HW performances evaluations due to the combination of multi-scale dense-prediction DNNs, heterogeneous HW targets, and deployment-based on manufacturer's tools.

In regard to these challenges, this paper proposes the following contributions:

- A practical automated HW-NAS framework including Nvidia' TensorRT black-box deployment tool through a co-design search on (1) DNN dense-prediction multi-branch topologies and (2) heterogeneous HW mappings compatible with the Nvidia Jetson Orin SoCs.

- A compact enhancement of the existing GPU/DLA mapping solution brought by TensorRT on the high-end embedded Jetson Orin series SoCs..
- A practical strategy to reduce a black-box DSE time bottleneck without losing accuracy with a hybridization of latency and power evaluations.
- Further analyses of obtained HW mappings for the Jetson Orin SoC series.

2 Related Work

In response to handcrafted design methods, HW-NAS became a better strategy by automatically handling the complex DSE optimization problem [3]. A typical HW-NAS flow is described in Fig. 1.

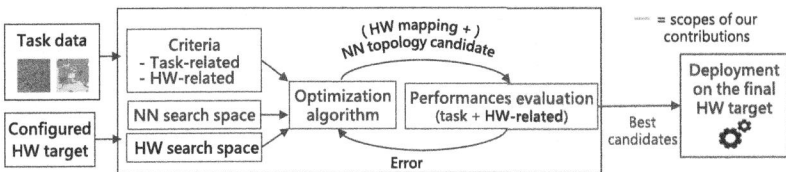

Fig. 1. HW-NAS flow standard components. *NN search space* refers to the available design hyper-parameters of the DNN architecture and *HW search space* to the HW target's inference configurations.

These methods are based on a *search algorithm* that handles the exploration of the design space. The optimization algorithm progressively selects the best candidates with respect to fixed criteria until the convergence towards an optimized neural architecture and HW settings solution. In order to converge towards an optimized result, current HW-NAS meet three main technical challenges: (1) the capability to explore a large enough search space in a reasonable time, (2) maintaining a high accuracy in the estimation of performance throughout the flow, and (3) leveraging the available hardware computing capabilities efficiently. At the end of this section, Table 1 lists the major HW-NAS propositions in the literature that, unlike our work, partially tackle these issues.

2.1 DSE Duration Issue

To limit the HW-NAS total time duration, all methods focus on exploring simple DNN architectures and/or a limited HW search space. For instance, [23,32] are considering either only the GPU or the CPU in heterogeneous SoCs, whereas all other references deal with *simple* DNN architectures, i.e. such as single branch or scale DNNs. Using simpler search spaces is convenient as it restricts the search complexity and so its exploration, but it limits the application performance or the energy efficiency and thus excludes some industrial use cases.

To ensure convergence in a reasonable amount of time (<4 d), a HW-NAS flow needs to perform precise and as fast as possible *hardware performances evaluations* (HW-PEs). If search spaces are small, relatively fast *HW-in-the-loop* (HIL) evaluations can be used as in [4]. But, if the HW measurements are slow and the search spaces not too complex ($< 10^{10}$ combinations or single branch), one may rely on faster performances estimators, like analytical models [6,12] or Deep *Look-Up-Tables*-based (LUT) models [7,16]. We refer to these estimators as *predictors* for the rest of this paper.

2.2 Accurate and Optimized DSE

In addition to new AI accelerators, manufacturers provide deployment tools that offer additional performance gains thanks to their expertise. For example, Nvidia's TensorRT deployment tool is now proposed with every new Nvidia's hardware product and is improved accordingly [18]. TensorRT's flow first modifies the DNN topologies, imposes a default HW mapping, and selects the fastest algorithms with respect to available HW resources. Then the optimized solution is compiled as a binary, ready for the inference execution. For a given DNN, only the use of TensorRT and its optimization process can favor the best inference performance on Nvidia's SoCs, as it acts as an internal DSE.

Nevertheless, the full use of manufacturer's tools is often discarded in related NAS works as it adds complexity or is time consuming. For instance, TensorRT's building phase can take up to 10 min, which is cumbersome in a NAS process where thousands of iterations are necessary. In addition, its optimizations based on black-box rules often prevent to accurately estimate inference latencies and power consumptions, hence prohibiting the use of fast performance predictors (see Sect. 3.1). Some works do consider deployment flows for multi-branch classification DNN within a reasonable time duration but, contrary to TensorRT, theses flows can be replaced with faster predictors, as they apply modelable or white-box optimizations (i.e. NN-stretch [27], NASGuard [25], Qin et al. [20]). For these reasons, NAS flows such as [5,19] use simplified hardware mappings which cannot reflect TensorRT's optimizations for multi-branch dense-prediction DNNs topologies, e.g. significant sequential and parallel layer fusions. Some others effectively use TensorRT, however they only perform on few [9,15] or very distinct [30] neural network architectures, limiting the design space exploration.

2.3 HW Implementation on Jetson Orin Series

Nvidia's Orin SoCs embed a CPU for general-purpose processing, an Ampere GP-GPU for general and intensive data parallelism, and fixed-function Deep Learning Accelerators v2 or v3 (DLA) for energy-efficient processing [18]. The GPU comprises CUDA cores for general operations and Tensor cores for specific convolutions. Its internal structure and behavior remain closed-source, like the DLA, but can be analyzed trough profiling tools [18]. These SoCs can infer a

DNN either on the GPU, on a single DLA, or on both with a *mixed* GPU-DLA mapping. By default, TensorRT uses mixed GPU-DLA mappings only if DLA-incompatibilities exist [18].

To enable accurate mixed GPU-DLA mappings and better leverage HW computing resources, certain works [9,19] modify the classic TensorRT deployment flow. This prevents the use of the *ONNX* definition, which adds uncertainties for mixed GPU-DLA mappings. Hence, a hand-made description taking as input both the DNN topology and the HW mapping can replace the *ONNX* one. It can use the TensorRT Python API interface.

By enabling exact layer-wise-mappings, the HW mapping DSE is enlarged with every possible combination of GPU-DLA mappings. However, not every mixed-mappings should be considered as it increases the complexity of the search space significantly. Other works reduce the search spaces with a limited but still significant number of HW mappings like layer-wise [9], kernel-wise [30] or medium-grained mappings [19,31]. Such mappings strategies may lead to feasible DSE for simple DNN search spaces, but would lead to intractable DSE when it comes to multi-branch dense-prediction DNNs.

Thus, all state-of-the-art solutions only partially answer the need of a practical multi-objective HW-NAS for new vision DNN architectures able to explore the design space including the Nvidia Jetson Orin SoC capabilities efficiently.

Table 1. Comparison of HW-NAS existing flows, using commercial tools or not.

Work	Explored search spaces		Deployment flow	
	NN topologies for dense predictions	Mappings on heterogeneous HW targets	Accurate HW measurements	Manufacturer's (black-box) optimizations
OFA [7], HELP [16], HW-EvRSNAS [22]			✓	
HADAS [4], Hendrickx et al. [12]				
RHNAS [1], GPUNet [26], CoNAX [6]			✓	✓
Map-and-Conquer [5], Magnas [19]		✓	✓	✓
NN-stretch [27], NASGuard [25], Qin et al. [20]				
MnasNet [23], RealtimeSeg [32]	✓		✓	✓
Our HW-NAS flow	✓	✓	✓	✓

3 Detailed Methodology

In this section, we describe our practical HW-NAS flow using a compact enhanced mixed HW search. In particular, we introduce the use of hybrid HW-PEs to tackle HW-PEs limitations.

Based on a handmade mixed mappings definition compatible with all possible 10^{55} DNN topologies [8], we define a new HW-NAS flow (Fig. 2) that, unlike previous works, benefits from the flexibility offered by the heterogeneous target (extra mixed-mappings) and the extra optimizations proposed by the manufacturer (through TensorRT). To limit the total DSE duration while benefiting from the heterogeneity, the allowed HW mappings rely on extreme coarse-grained partitions of the DNNs, i.e. only three mixed mappings are added for a single-branch DNN. An example of this mapping is fully detailed in Sect. 4.

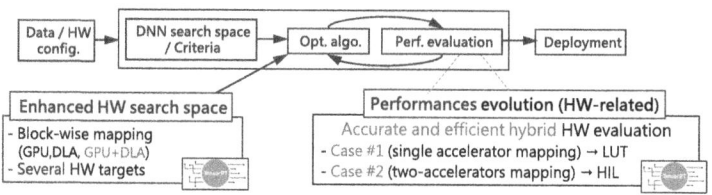

Fig. 2. Practical HW-NAS flow with hybrid evaluations and enhanced HW search space. The blue color indicates our contributions. (Color figure online)

3.1 Limitations of HW Performance Predictors

To be efficient without restricting the search spaces exploration, a HW-NAS flow has to use accurate and relatively fast HW-PEs. To ensure precision, a straightforward solution is to resort to HIL evaluations in the NAS flow. But it is not practical in our case since using the necessary TensorRT deployment flow is time-consuming. Another option is to rely on fast state-of-the-art predictors, but Orin SoCs are heterogeneous, closed-source and their programming depends on a closed-unpredictable optimization tool.

To identify compatible HW-PEs, we led a study using a single accelerator on the main three families of State-of-the-Art HW predictors compatible with multi-branch dense-prediction DNN topologies [8] and TensorRT's flow: LUTs, theoretical models based on the partially known HW features, and a State-of-the-Art regression model for non-linear deployment optimizations [30]. The results, reported in Table 2, show that the use of layer-wise LUTs is the only acceptable solution. The SoC theoretical models developed from Nvidia documentations [18], as well as the regression model turned out to be unusable solutions. LUTs obtained a mean relative error of 20 % over 200 latency predictions compared to actual measurements for complex DNNs, which is a tolerable error for our application. They also preserved the order between predicted values thanks to a correct correlation with a Kendall's tau of 0.7.

However, when applied to multiple accelerators, the LUTs become ineffective. For instance, we systematically observed important uncertainties for GPU-DLA mappings either regarding latency or power estimations. For the power, these uncertainties could result in a 85 % and 50 % latency and power errors, respectively. This prevents a HW-NAS convergence towards practical solutions. This phenomenon is explained by TensorRT's high sensitivity to operation changes

Table 2. Latency predictors study on the Orin AGX (MAXN mode). Result example for only one GPU (FP32). No predictor is 100% accurate.

Latency	Neural networks topologies	
predictors	*Correlation*	*Error*
Ideal model	1	5%
Look-Up Tables	0.7	+20%
Theoretical models		-80 to +325%
Regression models [30]	0.02 to 0.7	+44 to +180%

Accurate
Passable
Weak

between the DNNs topologies. Even with similar mappings, deviations occur because the use of resources differ according to operations and TensorRT optimizations.

These behaviors may not be limited to this study. Indeed, sources of uncertainty are common for heterogeneous SoCs, as they can be subject to runtime-level variability such as synchronization mechanisms (jitters) between the accelerators. To summarize, in our case and under certain conditions, usual HW-PEs can be used to accelerate the HW-NAS flow without degrading the exploration capacity.

3.2 Hybrid HW-PEs

The previous section shows that mappings with a single accelerator on Orin SoCs are the operating modes adapted for LUT predictors. From these observations, we introduce a practical strategy to use predictors when possible during HW-PEs. We define a *hybrid* evaluation scheme, i.e. relying on two paradigms, that uses fast predictors when they are accurate enough (i.e. with respect to the HW-NAS flow convergence and the application's constraints) and HIL evaluations otherwise (Fig. 3). In our case, LUT predictors are used for mappings with one accelerator (GPU or DLA) and HIL evaluations are used for mappings using the two accelerators. In this way, we favor a time reduction for the HW evaluation phase without compromising the flow convergence towards practical solutions.

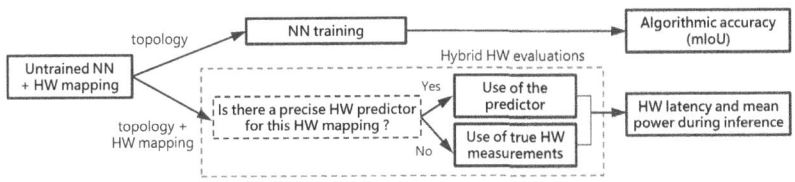

Fig. 3. Implemented hybrid HW-PEs strategy.

4 Experimental Setup

To demonstrate the performance of our HW-NAS flow, we selected a semantic segmentation problem for aerial detection and recognition applications. This kind of DNN can be typically embedded in satellites or recognition drones. As an illustration, we considered the Vaihingen dataset [21] from the Geoseg framework [24].

Then, the FasterSeg [8] search space was selected for its two-branch topology, capable of handling different information hierarchies and designed for semantic segmentation. Our HW-NAS flow relies on FasterSeg's original design option and on custom enhanced HW search spaces.

4.1 HW-NAS Flow Implementation

To explore the complex search space efficiently, the NAS flow uses multi-objective optimization on DNN and HW-related criteria with a genetic algorithm. With this kind of search algorithm, it is possible to manage the criteria independently and to deal more effectively with the non-convexity of the two search spaces. Still actively used in the HW-NAS domain [19], the genetic algorithm NSGA-II [10] is used to carry out the search multi-objective optimization with 1500 DNNs and HW mappings candidates on three criteria: the algorithmic accuracy (mIoU), the mean latency and the power consumption during inference on the HW target.

For HW-PE candidates, our flow relies on a hybrid evaluation strategy with HIL and LUT as HW-PEs, and we do not rely on FasterSeg's super-network for the training to preserve an accurate DNN performance estimation [29].

To limit the DSE duration, the NAS flow exploits three degrees of parallelism during the performance evaluation phase: at the training level through multi-GPU learning, between the training and the inference, and between hardware boards during HW-PEs. Thanks to this parallelism, we use full training to evaluate the DNN algorithmic accuracy without additional time penalty. The DNNs are trained in parallel on a 8 Nvidia A100 GPU cluster for 75 epochs. Trainings last up to 20 min for a group of 8 parallel DNN candidates, which is negligeable compared to HW performances retrieving (up to 100 min for 8 sequential DNN candidates).

The latency is obtained from TensorRT's *trtexec* tool and the mean instant power is read from three internal power rails (sum of VDD_GPU_SOC, VDD_CPU_CV, and VIN_SYS_5V0 rails) every 0.5 s [2]. Experiments are built with Pytorch 3.8, TensorRT 8.4 for the Jetson Orin AGX and TensorRT 8.5 for the Jetson Orin NX.

4.2 Search Spaces

The FasterSeg's search space focuses on a two-branch encoder followed by a fixed decoder. The encoder is composed of 16 successive operation blocks for each branch and is configurable trough two-level independent design options. A first macroscopic subspace determines the overall structure of the DNN with a set of increasing down-sample rates for each block. A second microscopic subspace determines the operations set and number of channels composing each block.

In order to ease the hardware mapping of the DNN, we define a block structure based on the original DNN structure of FasterSeg that we also use later for the hardware mappings (upper part of Fig. 4). By doing so, we limit the available mixed mappings to very distinct mappings with as few changes between accelerators as possible. Thus, for a typical DNN topology for image processing applications with b parallel branches, we propose the hardware mapping set depicted in the lower part of Fig. 4. Such a design favors additional performance trade-offs and allow removing less efficient mappings from the DSE as identified by authors of [2]. For instance, mappings with continuous sequences of operations on the same accelerator must be preferred to the others with systematic transfers between the GPU and the DLA.

Considering the HW targets search space, the NAS flow considers two representative heterogeneous SoCs from Nvidia's Jetson Orin series: the AGX Devkit 64 GB and the NX 16 GB, composed of the same AI accelerators, however with different numbers of cores and frequency, as explained in [18]. The Jetson Orin Nano is not included since it does not have any DLA. For a wider yet compact search space, the three default power modes for each Orin target are selected as additional HW hyper-parameters to be set outside the NAS flow. The full available search space is summarized Table 3.

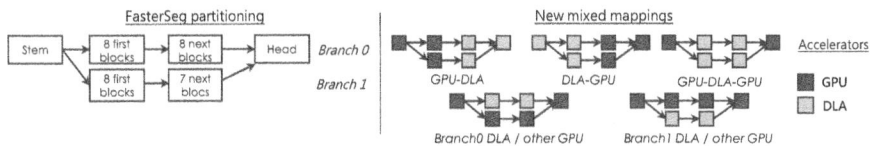

Fig. 4. Upper: Cuttings in FatserSeg NN structure for HW mapping. Lower: available HW mappings for FasterSeg, which has 2 parallel branches

Table 3. Available search spaces used for the experiments.

Search space	Design variables	Available values	# combinations
DNN (FasterSeg)	Downsample ratios	{8, 16, 32}	
	Operations	{identity, 3×3 conv, 2 3×3 conv, 3×3 zoomed conv, 2 3×3 zoomed conv}	10^{55} (calculated from [8])
	Channels	{4, 6, 8, 10, 12}	
HW	HW mappings[a]	{GPU, DLA, GPU-DLA-GPU, GPU-DLA, DLA-GPU} + {DLA on branch b_i, GPU on others}	7 (= 5+b)
	Target boards	{Orin AGX 64 GB}; {Orin NX 16 GB}	2
General settings	Power modes	{MAXN, 30 W, 15 W};{MAXN, 25 W, 10 W}	3

[a] TensorRT uses by default the CPU for data pre/post-processing

5 Results and Discussion

5.1 Impact on Space Exploration

Our HW-NAS flow's convergence result is presented in Fig. 5 where each dot represents an evaluated DNN topology and its associated HW mapping candidate. These solutions belong to the *Pareto Fronts* (PFs), i.e. the selected candidates and HW mappings offering the best trade-off for the 3 criteria after each step of the genetic algorithm. The general evolution of these PFs shows the benefits of a DSE within our configurations. For all criteria, better solutions are obtained through iterations compared to the initial solutions (dark blue points). In particular, as depicted in Fig. 6, the new HW space does not degrade the DNN algorithmic performance at the cost of better deployment-related performances: acceptable validation mIoU (>0.6) DNN solutions exist for all the HW mappings, either for low latency or low-power constraints.

In more details, Fig. 7 presents the HW space explorations, with the final full PF solutions (green dots) obtained from the different HW-NAS flows among all the explored intermediate full PF solutions (gray dots). Figure 7.a is the 2D projection of points from Fig. 5 on the HW criteria. To better identify the impact of introducing mixed HW mappings, the figure also depicts the projection for the HW criteria (latency and power consumption) of our final PF (dashed green line) and the default TensorRT PF (dashed black line). For the sake of clarity, the dashed lines suggest the 2D plan of these new HW PFs. The gap between our obtained HW-PF and TensorRT's HW PF highlights the benefit of the extended HW search space. Indeed, when it comes to the MAXN experiments, mixed mappings widen the explored HW space with new satisfying solutions.

The new PF is more evenly populated than for TensorRT (more diverse solutions) and the Hyper-Volume, i.e. the discovered space with regards to a reference point, is increased. On the other hand, TensorRT's default mappings (black-edged dots for GPU-only, DLA-only) offer the most extreme HW performances, either the lowest mean powers or the lowest latencies. As a result, it is possible with the new mixed mappings to have equivalent accurate DNN solutions with 50 % less power consumption or 80% less latency (with MAXN). With the new obtained PFs, it is now possible to choose among more diverse solutions to meet industrial constraints, like a maximal power budget.

5.2 Impact of Hybrid Evaluations

With hybrid HW-PEs applied to our experiments, significant time reduction is observed for some experiments with a limited impact on the space exploration or the flow convergence towards practical solutions, as shown in Table 4. Considering the MAXN power mode example, the total duration of the HW-NAS flow is reduced by 33 % compared to HIL evaluations. Besides, the observable Hyper-Volume loss is limited to -1 %. This is due to the use of inaccurate HW performance predictors for full-GPU and full-DLA mappings. Same trends are observed with the 10-15W power mode experiment. However, the duration is

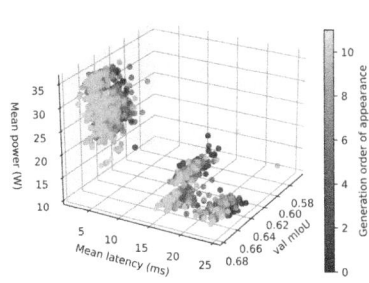
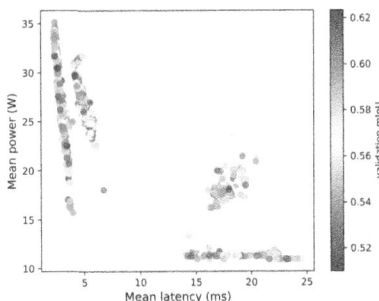

Fig. 5. Convergence of our HW-NAS flow among the tree criteria, with the successive Pareto front solutions obtained from NSGA-II (MAXN mode).

Fig. 6. Explored DNN performance (validation mIoU) from the obtained Pareto fronts (MAXN mode).

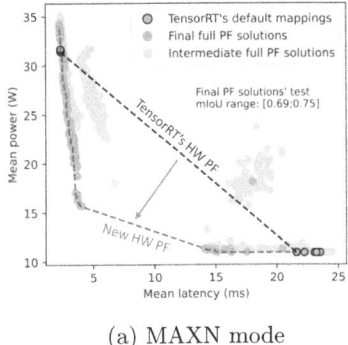
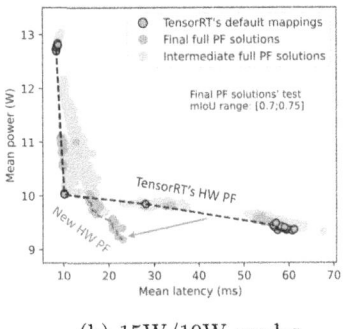

(a) MAXN mode

(b) 15W/10W modes

Fig. 7. Final Pareto Front (PF) solutions for all criteria (green dots) and Hardware-Pareto Front (HW PF, dashed lines) for each power mode experiment. The final PF solutions come from the *full* (Color figure online) 3D Pareto fronts (Fig. 5). The HW PF is built from PF solutions' 2D projection on the HW criteria (latency, power).

Table 4. Impact of hybrid HW-PEs

HW evaluations	Flow duration (h)	Accuracy interval (val mIoU)	Latency interval (ms)	Power interval (W)	Normalized Hyper-Volume[a]
HW-in-the-Loop	90	[62; 69]	[2.2; 24.7]	[10.9; 34.3]	0.81
Hybrid evaluations	62	[62; 69]	[2.2; 21.2]	[10.9; 33.3]	0.80
Difference	-33%	0%	−10%	−3%	-1%

[a] with reference point : (mIoU 0.5, 30 ms latency, 50W power)

only reduced by 9 % for the 25-30W power mode experiment. This results points out a limitation of our method: if a majority of mixed-mappings are selected, more HIL evaluations will be performed, resulting in a smaller time reduction but for the sake of convergence towards practical solutions.

5.3 Exploitation and Comparison with an Handcrafted NN

With the various obtained solutions from the NAS flow, it is now possible to select them accordingly to industrial constraints. To this aim, Table 5 details DNN performances for typical performances budget cases. For comparison, we selected BiseNetV2 [28], a lightweight handcrafted multi-branch CNN contemporary to FasterSeg. From these experiments, Table 5 highlights the flexibility of our HW-NAS solutions as they offer additional trade-offs compared to lightweight handcrafted DNNs. As an example, for a constrained number of parameters, power consumption or latency, the built HW-NAS takes more time for training and convergence but reduces the design time and provides solutions with a higher accuracy up to +4.4% test mIoU (+6% relative gain) or with lower power consumption (−30%).

Table 5. Comparison with a handcrafted DNN on the Vaihingen dataset on Jetson Orin AGX (MAXN) for selected budget constraints

NN	Constraints	Design time (days)	Training time (hours)	Accuracy (test mIoU)	Latency (ms)	Power (W)	# Params (millions)
BiseNetV2	Manual	> 100[a]	12.5	70.3	2.35	31.1	5.7
ours	high accuracy	1[b]	66	74.6	2.86	26.07	6.1
	<2.4 ms			74.5	2.36	31.61	8.5
	<31 W, <6 M			73.2	3.2	22.3	4.7

underline: constrained values, blue: better value than BiseNetV2
[a]: typical duration due to long trial/error steps, [b]: about 24 h are needed to build the LUTs

6 Conclusion and Future Work

Designing effective embedded AI-based applications for heterogeneous hardware platforms becomes increasingly difficult as the architectural complexity of DNNs increases. In this paper, we proposed a novel multi-objective and parallel HW-NAS flow able to leverage heterogeneous Jetson Orin SoCs, with a black-box proprietary deployment tool, and the FasterSeg's multi-branch architectures for semantic segmentation applications efficiently. Our flow preserves solutions' practicality while favoring efficiency thanks to hybrid evaluations of HW performances, considering the manufacturer's expertise, as well as a compact enhanced HW mapping search space based on TensorRT. By considering the FasterSeg framework, our HW-NAS flow is capable of offering, within a reasonable time (up to -33% duration compared to the inital flow), solutions optimized for different criteria that can be exploited according to the hardware or performance constraints of the application use case. Although the proposed strategy can be applied to every neural branch-based neural topologies, future works will focus on testing the generalization and revelance of this flow to other multi-branch multi-scale DNN architectures for dense tasks predictions (e.g. the Yolos family), different types of SoCs (e.g. Google's Pixel phone with CPU, GPU and NPU), and an in-depth comparison with FasterSeg's LUTs handled by its supernetwork.

References

1. Akhauri, et al.: RHNAS: realizable hardware and neural architecture search. arXiv preprint arXiv:2106.09180 (2021)
2. Archet, et al.: Energy-efficient use of an embedded heterogeneous SoC for the inference of CNNs. In: DSD, pp. 30–38 (2023)
3. Benmeziane, et al.: A comprehensive survey on hardware-aware neural architecture search. arXiv preprint arXiv:2101.09336 (2021)
4. Bouzidi, et al.: HADAS: hardware-aware dynamic neural architecture search for edge performance scaling. In: DATE, pp. 1–6 (2023)
5. Bouzidiand, et al.: Map-and-conquer: energy-efficient mapping of dynamic neural nets onto heterogeneous MPSoCs. In: DAC, pp. 1–6 (2023)
6. Braatz, et al.: CoNAX: towards comprehensive co-design neural architecture search using HW abstractions. In: ASAP, pp. 8–16 (2024)
7. Cai, et al.: Once-for-all: train one network and specialize it for efficient deployment. In: ICLR (2020)
8. Chen, et al.: FasterSeg: searching for faster real-time semantic segmentation. In: ICLR (2020)
9. Dagli, et al.: AxoNN: energy-aware execution of neural network inference on multi-accelerator heterogeneous SoCs. In: DAC, pp. 1069–1074 (2022)
10. Deb, et al.: A fast and elitist multiobjective genetic algorithm: NSGA-II. TEVC **6**(2), 182–197 (2002)
11. Hao, et al.: A brief survey on semantic segmentation with deep learning. Neurocomputing **406**, 302–321 (2020)
12. Hendrickx, et al.: Hardware-aware NAS by genetic optimisation with a design space exploration simulator. In: CVPR, pp. 2275–2283 (2023)
13. Holder, et al.: On efficient real-time semantic segmentation: a survey. arXiv preprint arXiv:2206.08605 (2022)
14. Jouppi, et al.: Motivation for and evaluation of the first tensor processing unit. Micro **38**(3), 10–19 (2018)
15. Kim, et al.: Energy-aware scenario-based mapping of deep learning applications onto heterogeneous processors under real-time constraints. Trans. Comput. **72**(6), 1666–1680 (2022)
16. Lee, et al.: HELP: hardware-adaptive efficient latency prediction for NAS via meta-learning. NeurIPS **34**, 27016–27028 (2021)
17. Liu, et al.: A survey on evolutionary neural architecture search. Trans. Neural Netw. Learn. Syst. **34**(2), 550–570 (2021)
18. NVIDIA: NVIDIA documentation (2024). https://www.nvidia.com
19. Odema, et al.: MaGNAS: a mapping-aware graph neural architecture search framework for heterogeneous MPSoC deployment. TECS **22**(5s), 1–26 (2023)
20. Qin, et al.: Searching tiny neural networks for deployment on embedded FPGA. In: AICAS, pp. 1–5 (2023)
21. Rottensteiner, et al.: ISPRS semantic labeling contest. ISPRS: Leopoldshöhe, Germany **1**, 4 (2014)
22. Sinha, et al.: Hardware aware evolutionary neural architecture search using representation similarity metric. In: WACV, pp. 2628–2637 (2024)
23. Tan, et al.: MnasNet: platform-aware neural architecture search for mobile. In: CVPR, pp. 2820–2828 (2019)
24. Wang: Geoseg GitHub repository. https://github.com/WangLibo1995/GeoSeg

25. Wang, et al.: NASGuard: a novel accelerator architecture for robust neural architecture search (NAS) networks. In: ISCA, pp. 776–789 (2021)
26. Wang, et al.: Searching the deployable convolution neural networks for GPUs. In: CVPR, pp. 12227–12236 (2022)
27. Wei, et al.: NN-stretch: automatic neural network branching for parallel inference on heterogeneous multi-processors. In: MobiSys, pp. 70–83 (2023)
28. Yu, et al.: BiSeNet V2: bilateral network with guided aggregation for real-time semantic segmentation. IJCV **129**, 3051–3068 (2021)
29. Zhang, et al.: DCNAS: densely connected neural architecture search for semantic image segmentation. In: CVPR, pp. 13956–13967 (2021)
30. Zhang, et al.: NN-meter: towards accurate latency prediction of deep-learning model inference on diverse edge devices. In: MobiSys, pp. 81–93 (2021)
31. Zhou, et al.: Multi-accelerator neural network inference via TensorRT in heterogeneous embedded systems. In: COMPSAC, pp. 463–472 (2024)
32. ZiWen, et al.: Multi-objective neural architecture search for efficient and fast semantic segmentation on edge. Trans-IV **9**(1), 1346–1357 (2023)

Endoscopy Image Classification for Wireless Capsules with CNNs on Microcontroller-Based Platforms

Paola Busia[1](✉), Andrea Pinna[2], and Paolo Meloni[1]

[1] DIEE, University of Cagliari, 09123 Cagliari, Italy
{paola.busia,paolo.meloni}@unica.it
[2] Sorbonne University, CNRS, LIP6, 75005 Paris, France
andrea.pinna@lip6.fr

Abstract. Wireless Capsule Endoscopy (WCE) offers an important diagnostic instrument for different gastrointestinal diseases. Enhancing the WCE device with real-time image processing capabilities allows to assist specialized physicians in the long and cumbersome process of inspecting the significant amount of data acquired during the examination procedure, providing the first detection of the signs of relevant diseases that require further attention. In this work, we evaluate different state-of-the-art Convolutional Neural Network models for real-time WCE image classification, focusing on lightweight topologies suitable for execution on low-power microcontroller platforms and integration on the WCE device. The selected WCE-SqueezeNet model achieves 98.5% accuracy in the classification of ulcerative colitis, polyps, and esophagitis against healthy samples, allowing classification at a 16 fps rate on the GAP9 multi-core platform, with 61 ms inference time and 30.6 mW average core power consumption.

Keywords: Wireless Capsule Endoscopy · Near-Sensor Processing · Convolutional Neural Networks

1 Introduction

Gastrointestinal (GI) diseases represent a relevant concern for the health of millions of people worldwide. As a reference, the number of patients living with a GI condition in Europe between 2000 and 2019 was estimated to be over 332 millions [1]. Wireless Capsule Endoscopy (WCE) represents a common diagnostic instrument for the early detection of GI diseases, which enables proper medical intervention before more serious complications arise.

The current examination procedure involves image acquisition with the WCE device along the GI tract, allowing the detailed exploration of the tissue, and the transmission of a huge amount of image data to an external server for direct medical examination. Image processing and classification of common conditions through Artificial Intelligence (AI) approaches have been evaluated in the literature, with the aim of assisting the physicians in the diagnostic process [2,8,10]. Nonetheless, the continuous stream of images to the server through the wireless channel requires a significant amount of bandwidth, thus in-place processing has been explored [11], based on the general AI-at-the-edge trend in medical applications [3].

This work focuses on enabling near-sensor image processing capabilities on the WCE device, in order to perform real-time image classification and limit the WCE-to-server transmission only to the images representing a recognized symptom of the disease. To this aim, we target image classification based on Convolutional Neural Networks (CNNs) with a complexity suitable for efficient inference execution on low-power microcontroller-based platforms, that are compatible with the integration on the WCE device. Recent AI-oriented platforms in the edge domain leverage up to a few MBs of available memory [13], although the working memory of most common microcontroller-based platforms is within 1 MB. The targeted computing platforms thus introduce a significant constraint on the complexity of the classification model, both in terms of memory requirements and the number of required operations to ensure real-time processing.

The contributions of this work can thus be summarized in two main points:

- the evaluation on an open-source dataset of a suitable CNN classifier, the WCE-SqueezeNet model, for real-time near-sensor WCE image classification, reaching 98.5% accuracy in the recognition of three common GI conditions, including the assessment of the most effective image resolution reduction as a trade-off between accuracy and computational complexity;
- the preparation, deployment, and demonstration of real-time execution on the GAP9 low-power multi-core platform, enabling a 16 fps throughput within a core power envelope of 30.6 mW.

The paper is organized as follows: Sect. 2 summarizes the state of the art, Sect. 3 describes our proposed approach for the classification model training and evaluation, Sect. 4 reports the experimental results, and finally Sect. 5 summarizes the conclusions.

2 Related Work

The use of AI, particularly of CNN models, for medical image processing is well documented in the literature. Table 1 summarizes recent works from the state of the art, addressing WCE-image classification with CNN models, and referencing the same set of open source data, with the exception of [11].

The author of [8] presents a classification model obtained from the combination of truncated versions of the EfficientNetB0, MobileNetV2, and ResNet50V2 topologies, exploited as feature extractors prior to a Fusion Residual Block producing classification. The classification model, called MFuRE-CNN, reaches 97.75% accuracy in the recognition of 3 pathological conditions against healthy samples.

The EfficientNet topology is also exploited in the work of [10], where the EfficientNetV2B2 topology is adapted to the WCE task with the integration of a custom classification head including Global Average Pooling and Dense layers. The finally obtained GastroNet model outperforms the evaluated alternatives fine-tuned from state-of-the-art topologies, such as ResNet50, and EfficientNetv2B1, reaching over 99% accuracy.

Additionally, the work of [2] presents the DCDS-Net model, exploiting several blocks of separable convolutions prior to a classification block composed of three Dense layers.

As can be noticed from the table, the efficiency of the classifiers described in these works was not assessed on an embedded hardware target, however, due to their storage requirements, of at least 20 MB, and the number of operations required per inference run, over 2 GOPS, they do not represent suitable candidates for integration in an intelligent WCE device. On the other hand, the idea of exploiting network models reaching state-of-the-art accuracy on the ImageNet dataset was demonstrated as a successful approach, to be considered also for WCE classification.

A system based on real-time image processing on the WCE device is envisioned and assessed in the work of [11], where memory constraints and efficiency are taken into account for the selection of the CNN detection model, showing 99.5% average precision in the recognition and detection of colorectal polyps. The precision number refers to a 25% intersection-over-union between the detected bounding box and the ground truth, evaluated on the data acquired from 255 patients of Denmark's national screening program. The number of parameters is still significantly high, over 3 million, but it is compatible with the proposed camera-pill hardware architecture, integrating 8 MB of memory, and where the average power consumption was assessed to be around 50 mW.

In this work, we target a similar problem, aiming at real-time classification of different GI diseases. Exploiting the feature extraction capabilities, derived from learning on large image datasets, of the pre-trained state-of-the-art SqueezeNet

Table 1. Comparison with the state of the art of WCE image classification models.

Model	Accuracy	Precision	Recall	Deployed	Parameters Memory	GOPS*
MFuRE-CNN [8]	97.75%	97.75%	97.75%	✗	19.2 MB	7.8
ResNet50 [10]	98%	98.1%	98%	✗	89 MB**	7.6**
EfficientNetV2B1 [10]	98.5%	98.5%	98.5%	✗	25.9 MB**	2**
GastroNetV1 [10]	99.2%	99.3%	99.3%	✗	32.8 MB**	2.3**
DCDS-Net [2]	99.33%	99.37%	99.32%	✗	83.8 MB	9**
YOLO-based [11]	99.5 AP	N.R.	N.R.	✓	3.2 MB	0.9**
this work	98.5%	98.6%	98.5%	✓	750 kB	0.45

* The number of operations refers separately to multiplications and additions: 1 MAC = 2 OPS.
** Estimated from paper.

network, our proposed WCE-SqueezeNet classification model reaches a competitive accuracy compared to the alternatives, within a complexity and memory footprint suitable for inference deployment on resource-constrained low-power hardware platforms, demonstrated through direct measurements. The efficiency of the proposed model surpasses the topology presented in [11], both in terms of required parameters and operations.

3 Method

This section describes our training and evaluation approach for the development of the proposed classification system, presenting the reference dataset and target hardware platform, as well as the CNN model considered for the assessment.

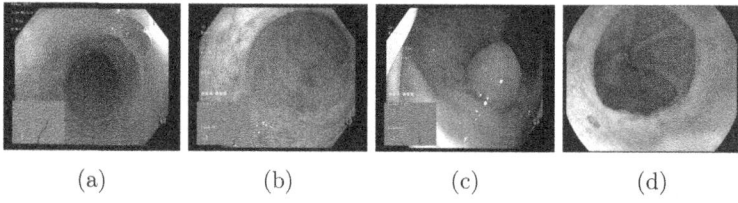

Fig. 1. Sample images from the reference dataset [9,14], including a) normal sample, b) ulcerative colitis, c) polyp, d) esophagitis.

3.1 Dataset

This study references the KVASIR [9] and the ETIS-Larib Polyp [14] databases, according to the data organization introduced in [8]. This collection counts 6000 images acquired through WCE, including an equal number of examples for three main pathological conditions of the gastrointestinal tract, such as ulcerative colitis, polyps, and esophagitis, as well as healthy/normal samples. Figure 1 reports an example of acquired image for each of the targeted classes.

Reference [8] also introduced a standard training, validation, and test split, according to the scheme summarized in Table 2. All the images in the dataset were pre-processed in order to standardize their size to a 224×224 resolution and normalized according to the data format expected by the different models considered. The selection of the input resolution was then the subject of a dedicated exploration, which is described in detail in Sect. 4.

Table 2. Data organization into training, validation, and test set.

Class	Train	Valid	Test
Normal - N	800	500	200
Ulcer - U	800	500	200
Polyps - P	800	500	200
Esophagitis - E	800	500	200
Tot	3200	2000	800

3.2 Hardware Target

To evaluate the efficiency of the WCE image classification application, we consider as a target the GAP9 Parallel Ultra-Low-Power (PULP) platform [4]. This device recently demonstrated remarkable energy efficiency in the tiny-ML benchmarks [7], with 0.033mW/GOP. It is an advanced microcontroller-based platform, integrating a cluster of nine parallel processors, which have access to a shared 128 kB L1 memory. The cluster can be exploited for parallel processing and the acceleration of typical deep learning workloads, such as convolutional and fully connected layers. The memory hierarchy also includes a 1.5 MB L2 memory, thus providing enough storage and computational resources to accommodate the classification model, within the limited power budget compatible with integration on WCE devices.

3.3 Classification Approach

A common approach in medical image classification problems is to leverage the feature extraction capabilities of off-the-shelf models pre-trained on large image datasets, such as ImageNet, and finally specialize them for the task at hand [6]. This solution often results in higher classification performance than training the same topology from scratch, as the available medical data is typically reduced and sometimes unbalanced in the representation of the different conditions.

Considering these documented results, in this work we aim to fine-tune an image classification model for WCE image classification. The network topology was selected based on the assessment of the computational and storage requirements, with the aim of targeting ultra-low-power deployment on tiny microcontroller-based platforms, to perform near-sensor image processing directly on the WCE device. We thus defined a memory constraint of 1 MB as the maximum acceptable memory footprint, limiting the evaluation to network models exploiting less than 1 million parameters. Therefore, we selected the SqueezeNet topology [5] as the backbone of our classification model. The structure of the model is recalled in Fig. 2. Compared to the Vanilla topology, we replaced the classification head with a dense layer suitable for the new 4-classes problem. In this configuration, the model exploits less than 740k parameters, thus allowing to meet the memory constraint with 8-bit quantization.

For the fine-tuning of the model, we leveraged the Pytorch framework, exploiting CrossEntropy loss, SGD optimizer, and 0.001 learning rate. The learning rate was iteratively adapted after patience of 30 epochs without improvements on the validation set, considering up to 6 steps. The final update was considered as the early stop condition. The training was performed on Google Colaboratory, leveraging the T4 GPU.

4 Experimental Results

In this section, we summarize the experimental results obtained on the target dataset, highlighting the most relevant metrics describing the classification performance, and including the impact of quantization on the final accuracy. We finally assess the efficiency of the selected solution with on-hardware direct measurements, demonstrating the feasibility of performing real-time classification on suitable low-power hardware targets.

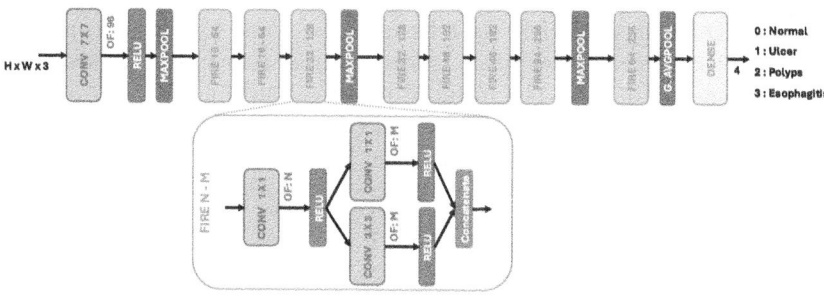

Fig. 2. WCE-SqueezeNet classification model.

4.1 Classification Performance Assessment

As the first step of the performance investigation, we compared the achievable accuracy when training and testing the classification model on images of different resolutions, starting from the 224×224 resolution exploited in [8], then reducing the size of the training image to 128×128, and finally to 64×64. The outcome of the exploration is summarized in Fig. 3, where the evaluated alternatives are placed according to the accuracy achieved on the validation set and to their computational complexity in terms of number of required operations (GOPS). Each model in the plot was evaluated on images of the same resolutions as the examples learned during the training. As can be observed, reducing the image size to 128×128 introduces only a negligible drop in the accuracy, while resulting in a significant reduction, by a factor of 3×, of the computational workload. On the contrary, the performance degrades significantly when the resolution is reduced further.

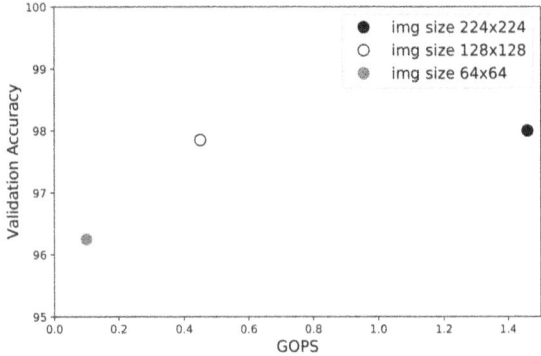

Fig. 3. Exploration of the input image resolution considering the validation set.

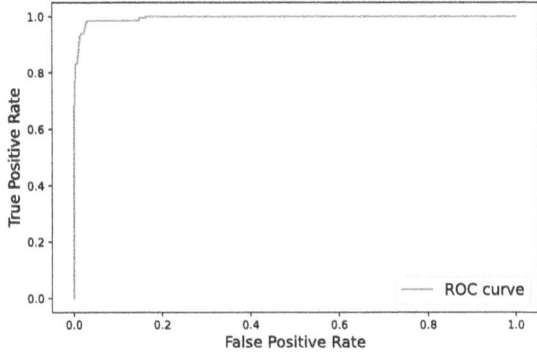

Fig. 4. Receiver Operating Characteristic curve for polyps recognition.

Based on this first assessment, and considering the low-power hardware targeted for the deployment, we selected the model trained to perform classification on the 128×128 images. We then performed a refinement training, based on the exploration of the most relevant hyperparameters, with batch size equal to 128, reaching an accuracy of 99% for full precision inference on the test set.

The details of the confusion matrix obtained with the WCE-SqueezeNet model are reported in Table 3a. As can be observed, the model shows perfect specificity, providing 100% recall in the recognition of the normal condition, with no false alarms raised based on the examples in the test set. The average precision and recall on all the targeted classes are 99%. Due to the importance of polyps' early detection, we further explored the accuracy in the recognition of this target class. Figure 4 shows the Receiver Operating Curve for the recognition of the polyps class against all other classes in the dataset. As can be noticed, the model provides a good discrimination ability, with an area under the curve (AUC) value equal to 0.99.

Table 3. Confusion Matrix resulting from test set classification with the WCE-Squeezenet model.

(a) Full Precision.

		N	U	P	E
True Labels	Normal	200	0	0	0
	Ulcer	0	198	2	0
	Polyps	1	4	195	0
	Esophagitis	0	1	0	199

Predicted Labels

(b) 8-bit Integer.

		N	U	P	E
True Labels	Normal	200	0	0	0
	Ulcer	0	197	3	0
	Polyps	3	4	193	0
	Esophagitis	0	1	1	198

Predicted Labels

4.2 Quantization

As anticipated, the aim of this work is to enable real-time classification on the WCE device, targeting execution on low-power microcontrollers. In order to meet the memory constraint of 1 MB, the memory requirements of the WCE-SqueezeNet model needed to be reduced through quantization to 8-bit precision. The quantization was performed through the TensorflowLite utilities, resulting in a limited accuracy drop, to 98.5%, compared to the Floating Point 32-bit full precision representation. The confusion matrix obtained on the test set with the integer model is reported in Table 3b. As can be noticed, only a few errors involving the pathological classes were introduced.

4.3 Discussion

Table 4. Classification Performance Assessment on the test set. PT column indicates whether the training started from ImageNet trained weights.

Model	PT	Accuracy	Ulcer		Polyps		Esophagitis	
			Precision	Recall	Precision	Recall	Precision	Recall
WCE-Squeezenet	✓	**98.13**	98.45	95	95.12	97.5	100	100
WCE-Squeezenet	✗	84.63	64.84	88.5	81.97	50	98	100
Squeezenet + SVM	✓	96.5	95	95.5	95.29	91	100	100
WCE-MobileNetV2	✓	**98.5**	97	97	97	97	100	100
MobileNetV2 + SVM	✓	96.87	96.39	93.5	94	94	100	100

In this section, we discuss the effectiveness of the classification approach described in Sect. 3. First, we evaluate the training approach, comparing it to the classification performance achievable with:

– the same topology trained from scratch, with no previous knowledge acquired on the ImageNet dataset;

– the pre-trained feature-extraction model, combined with a classifier trained on the target problem.

Additionally, we also considered the comparison with a more complex network model, such as MobileNetV2 [12]. Table 4 summarizes the comparison, based on models trained on images of 224×224 resolution. As can be observed, standalone CNN classification outperforms the classification of the extracted features based on SVM for both the network models considered. Furthermore, starting from the pre-trained parameters provides a significant advantage over training the same topology from scratch. Finally, the classification performance enabled by the WCE-SqueezeNet model is very close to the best one achievable with the more complex MobileNet topology.

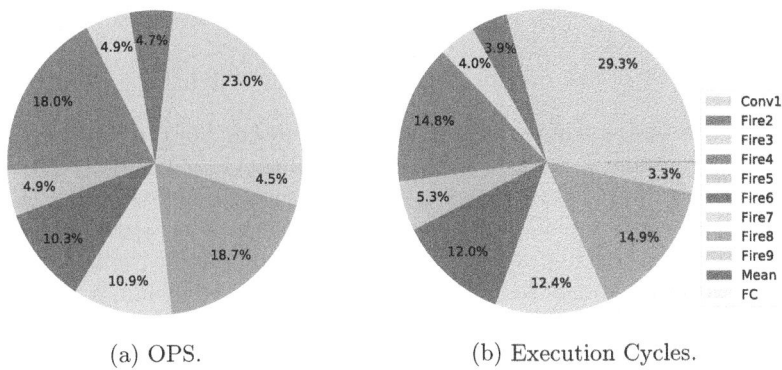

(a) OPS.   (b) Execution Cycles.

Fig. 5. Computational workload of the different layers in WCE-SqueezeNet model evaluated in terms of number of required operations and required execution cycles on the GAP9 platform, for 224×224 input resolution.

4.4 Deployment

The deployment of the selected WCE-SqueezeNet model was automated through the proprietary code generation tool, the GAP9 SDK. We compared the required inference time for the model applied to 224×224 and to 128×128 input images, considering parallel execution on eight cores of the computing cluster. In the first case, inference time was measured equal to 0.2 s, thus resulting in an expected throughput of 5 fps, evaluated at a 370 MHz working frequency. The average computational efficiency was 9 OPS/cycle. In the second case, the reduced computational workload resulted in only 61 ms inference time, with an expected throughput of 16 fps and an average computational efficiency of 10 OPS/cycle.

The composition of the computational workload is reported in Fig. 5a and 6a, whereas the required inference time on the GAP9 platform for each layer is represented in Fig. 5b and 6b. As can be noticed, the composition of the expected

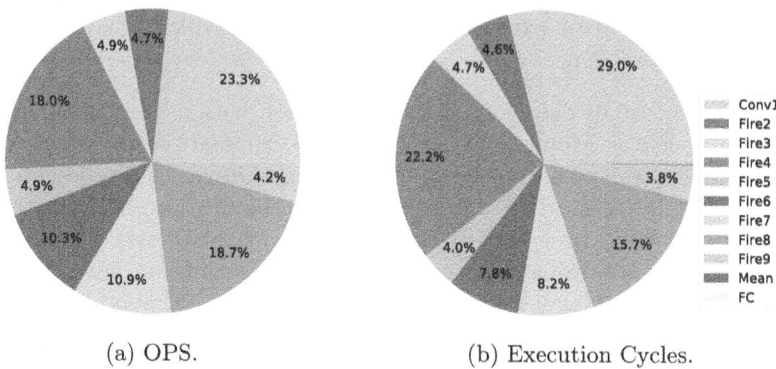

(a) OPS.   (b) Execution Cycles.

Fig. 6. Computational workload of the different layers in WCE-SqueezeNet model evaluated in terms of number of required operations and required execution cycles on the GAP9 platform, for 128×128 input resolution.

workload based on the number of required operations and of the measured inference time is very similar, showing there is no significant inefficiency in the implementation of the most relevant operands. The computational workload is dominated by the convolutional layers, as the main operands exploited in the Fire modules, while the contribution of the fully connected classification head is negligible. Input resolution shows only a limited impact on the composition of the required execution time.

Finally, we assessed the energy efficiency, by measuring the average core power consumption during inference execution, which is equal to 30.6 mW. The required energy per inference is thus 1.9 mJ. Measurements were performed with Nordic Power Profiler II. This result demonstrates the suitability of performing real-time inference on the WCE device, with limited power requirements compatible with battery-powered solutions.

5 Conclusions

In this work, we presented a classification model for the recognition of GI diseases based on WCE acquisition. The WCE-SqueezeNet model demonstrated 98.5% classification accuracy, evaluated after 8-bit quantization for efficient inference execution on the targeted GAP9 low-power platform. The analysis of the accuracy degradation with the progressive reduction of the input image resolution showed that compression up to a 128×128 resolution is possible with a negligible impact on the accuracy. The efficiency of the proposed solution was evaluated on the GAP9 platform, considering parallel execution on 8 cores. The measurements demonstrated a 16 fps achievable throughput and an average core power consumption of 30.6 mW, compatible with possible integration in the WCE device.

Acknowledgments. We acknowledge financial support under the National Recovery and Resilience Plan (NRRP), Mission 4 Component 2 Investment 1.5 - Call for tender No.3277 published on December 30, 2021 by the Italian Ministry of University and Research (MUR) funded by the European Union – NextGenerationEU. Project Code ECS0000038 – Project Title eINS Ecosystem of Innovation for Next Generation Sardinia – CUP F53C22000430001- Grant Assignment Decree No. 1056 adopted on June 23, 2022 by the Italian Ministry of University and Research (MUR). This research was also supported by the French National Research Agency (ANR) under the LabCom program 2021 - V2 (ICI-Lab), by the European Union's Horizon 2020 Research and Innovation Program under Grant Agreement GA 101140052 (H2TRAIN), and by NextGenerationEU Mission 4, Component 2, Investment 1.5, CUP B83C22002820006—Project METBIOTEL—Innovation Ecosystem ECS 0000024 ROME TECHNOPOLE SPOKE 1, and SPOKE 6.

References

1. Tackling the burden of digestive disorders in Europe. The Lancet Gastroenterol. Hepatol. **8**, 95 (2023). https://doi.org/10.1016/S2468-1253(22)00431-9
2. Asif, S., Zhao, M., Tang, F., Zhu, Y.: DCDS-Net: deep transfer network based on depth-wise separable convolution with residual connection for diagnosing gastrointestinal diseases. Biomedical Signal Processing and Control **90**, 105866 (2024). https://doi.org/10.1016/j.bspc.2023.105866, https://www.sciencedirect.com/science/article/pii/S1746809423012995
3. Busia, P., Scrugli, M.A., Jung, V.J.B., Benini, L., Meloni, P.: A tiny transformer for low-power arrhythmia classification on microcontrollers. IEEE Trans. Biomedical Circuits Syst., pp. 1–11 (2024). https://doi.org/10.1109/TBCAS.2024.3401858
4. GreenWaves: ultra low power gap processors (2024). https://greenwaves-technologies.com/low-power-processor/
5. Iandola, F.N., Han, S., Moskewicz, M.W., Ashraf, K., Dally, W.J., Keutzer, K.: SqueezeNet: alexnet-level accuracy with 50x fewer parameters and <0.5 mb model size. arXiv:1602.07360 (2016)
6. Li, X., Cen, M., Xu, J., Zhang, H., Xu, X.S.: Improving feature extraction from histopathological images through a fine-tuning imagenet model. J. Pathol. Inform. **13**, 100115 (2022). https://doi.org/10.1016/j.jpi.2022.100115, https://www.sciencedirect.com/science/article/pii/S215335392200709X
7. ML Commons: inference: tiny. v1.0 Results (2024). https://mlcommons.org/en/inference-tiny-10/. Accessed 30 Oct 2024
8. Montalbo, F.J.P.: Diagnosing gastrointestinal diseases from endoscopy images through a multi-fused CNN with auxiliary layers, alpha dropouts, and a fusion residual block. Biomedical Signal Processing and Control **76**, 103683 (2022). https://doi.org/10.1016/j.bspc.2022.103683, https://www.sciencedirect.com/science/article/pii/S1746809422002051
9. Pogorelov, K., et al.: KVASIR: a multi-class image dataset for computer aided gastrointestinal disease detection. In: Proceedings of the 8th ACM on Multimedia Systems Conference, pp. 164-169. MMSys'17, Association for Computing Machinery, New York, NY, USA (2017). https://doi.org/10.1145/3083187.3083212
10. Rajkumar, S., et al.: GastroNet: a CNN based system for detection of abnormalities in gastrointestinal tract from wireless capsule endoscopy images. AIP Adv. **14**(8), 085223 (2024). https://doi.org/10.1063/5.0208691, https://doi.org/10.1063/5.0208691

11. Sahafi, A., et al.: Edge artificial intelligence wireless video capsule endoscopy. Sci. Rep. **12**, 13723 (2022). https://doi.org/10.1038/s41598-022-17502-7, https://doi.org/10.1038/s41598-022-17502-7
12. Sandler, M., Howard, A., Zhu, M., Zhmoginov, A., Chen, L.C.: MobileNetV2: inverted residuals and linear bottlenecks. In: Proceedings of the IEEE Conference on Computer Vision and Pattern Recognition, pp. 4510–4520 (2018)
13. Scherer, M., et al.: Deeploy: enabling energy-efficient deployment of small language models on heterogeneous microcontrollers. IEEE Trans. Comput. Aided Des. Integr. Circuits Syst. **43**(11), 4009–4020 (2024). https://doi.org/10.1109/TCAD.2024.3443718
14. Silva, J., Histace, A., Romain, O., Dray, X., Granado, B.: Toward embedded detection of polyps in WCE images for early diagnosis of colorectal cancer. **9**, 283–293 (2014). https://doi.org/10.1007/s11548-013-0926-3

Joint Underwater Depth Estimation and Dehazing from a Single Image Using Attention U-Net

Saqib Nazir[1]( ), Reza Mohammadi Asiyabi[2], and Olivier Lezoray[1]

[1] Normandie Univ, UNICAEN, ENSICAEN, CNRS, GREYC, Caen, France
saqib.nazir@unicaen.fr
[2] School of GeoSciences, The University of Edinburgh, Edinburgh, UK
reza.asiyabi@ed.ac.uk

Abstract. Underwater imaging presents unique challenges compared to open-air photography, primarily due to diminished visibility and geometric distortions, impeding the development of underwater Computer Vision (CV) and robotic vision perception. Previous methods relying on simplified image formation models for image enhancement have often yielded unsatisfactory results. This paper proposes a new deep learning-based architecture for joint depth estimation and dehazing from a single underwater monocular image, seeking to take advantage of the mutual benefits between these two interrelated tasks. The proposed architecture is a Two-Headed Depth Estimation and Dehazing Attention Network (2HDED:AttN) with an end-to-end training approach. Comprehensive experiments on synthetic and real underwater datasets showcase the proposed architecture's superior performance in jointly addressing underwater depth estimation and image dehazing tasks. The method effectively estimates underwater depth and improves underwater image quality, paving the way for enhanced underwater computer and robotic vision applications.

Keywords: Underwater Image Enhancement · Depth estimation · Image dehazing · Multi-task learning

1 Introduction

Underwater imaging is an interesting challenge due to the increasing number of applications in ocean research such as offshore oil exploitation and deploying and maintaining underwater structures such as pipelines and cables. The main challenge in underwater exploration lies in 3D reconstruction, which necessitates accurate depth or 3D point estimation. However, underwater imaging is significantly hindered by optical distortions, including haze, resulting in degraded object perception and visibility. Additionally, color and contrast are severely compromised in underwater images, due to light scattering and absorption, limiting their effectiveness for visual surveys [26]. Accurately dehazed images can significantly improve underwater applications' automatic segmentation, feature-matching accuracy, and navigation capabilities.

Traditional techniques for underwater depth estimation have primarily relied on specialized hardware sensors, such as laser scanners, Time-of-Flight (ToF) cameras, or structured light systems [26]. These sensors offer accurate depth measurements but are often expensive, bulky, and have limited underwater range. To address these limitations, image-based depth estimation methods using single or multiple images are explored to infer depth information. These methods offer portability and flexibility but often suffer from accuracy issues due to the inherent challenges of underwater imaging, such as light attenuation and scattering.

Early research on underwater depth estimation was based on traditional CV techniques and optical principles [12]. These methods employ image analysis to extract edge directions, frequency coefficients, and depth-related features in specific image segments [2] or Global features representing the overall scene [15]. Previous studies considered global and local structural similarities within the scene, followed by optimization to enhance depth estimation accuracy [5]. However, the inherent limitations of monocular cameras still lead to a loss of scale when recovering scene depth from RGB images.

Recent advancements in CV and deep learning have opened up new possibilities for monocular depth estimation in underwater environments [5,17]. Deep learning models have demonstrated remarkable capabilities in estimating depth from a single open-air image [14]. These models exploit the rich information contained in images to infer depth cues, overcoming the limitations of traditional hardware-based methods. The notable success of deep learning-based methods for depth estimation from open-air images conveys the potential of employing deep learning-based models for underwater depth estimation [21]. Deep learning models rely on extensive data for effective training. Due to the challenges of obtaining underwater ground truth data, training images are often synthetically generated [25]. Recently, several studies have introduced datasets comprising real and synthetic underwater imagery which facilitated the training of deep learning methods for underwater applications [2,7].

In this study, we propose the 2HDED:AttN architecture for jointly estimating underwater depth and dehazing the RGB image from a single hazy image. 2HDED:AttN is built on a successful deep learning architecture for open-air depth estimation and image deblurring proposed in [14]. Despite the remarkable performance of [14] model for open-air depth estimation, it struggles to generate high-quality results for underwater scenes, due to the higher complexity and optical distortions. Several enhancements, including a deeper encoder and attention module, are made in 2HDED:AttN to increase its ability to capture complex features and hierarchical representations of underwater scenes. The main contributions of this study are as follows:

- A new lightweight two-headed deep learning architecture is proposed for simultaneous depth estimation and RGB image dehazing for underwater images.

- An important feature of 2HDED:AttN is that once the model is fully trained, we can perform a task when the other head is removed, e.g. we can perform depth estimation without the dehazing decoder and vice versa.
- Comprehensive qualitative and quantitative analyses applied to the proposed architecture against state-of-the-art underwater depth estimation and image dehazing methods.

2 Related Work

Various methods have been proposed for underwater depth estimation in the literature. Campos et al. [3] give an overview of various sensors and methodologies used for underwater depth estimation. The Dark Channel Prior (DCP) was the first method proposed by [6], for single-image haze removal using depth estimation. Later, multiple variations of DCP have been proposed for underwater depth estimation and image dehazing [3]. Chen et al. [2] introduced an Encoder-Decoder model, leveraging the Laplacian pyramid-based decomposition technique in decoding to proficiently restore depth boundaries and enhance the overall distribution of depth maps.

Traditional underwater image enhancement methods can be categorized into image-based and model-based algorithms. Image-based methods [1,21] estimate the transmission map of underwater images to remove the haze and apply the color correction based on this map. On the other hand, model-based methods [24], model the underwater imaging process with respect to the underwater optical properties to recover the dehazed image. With its powerful learning capacity, deep learning presents promising solutions for open-air image dehazing, which can be transferred for underwater image enhancement applications [22]. A convolutional architecture is proposed in [19] to estimate ambient light and transmission map. Wang et al. [20] proposed an end-to-end framework for jointly inferring transmission map and correcting the color images. Li et al. [10] introduced an extra structural similarity index measure loss under the backbone of cycleGAN to correct the color in an unsupervised manner. Espinosa et al. [4] proposed an efficient deep learning architecture using discrete wavelet transform skip connections and channel attention modules for underwater image enhancement focused on solving key image degradations related to blur, haze, and color casts and inference efficiency. WaterGAN [9] generated underwater images from in-air images and depth pairs in an unsupervised pipeline, and trained a SegNet on the synthetic dataset for image correction.

3 Proposed Method

3.1 2HDED:AttN Architecture

Figure 1 demonstrates the proposed 2HDED:AttN architecture, built on a recently introduced network in [13] for estimating depth from defocused open-air

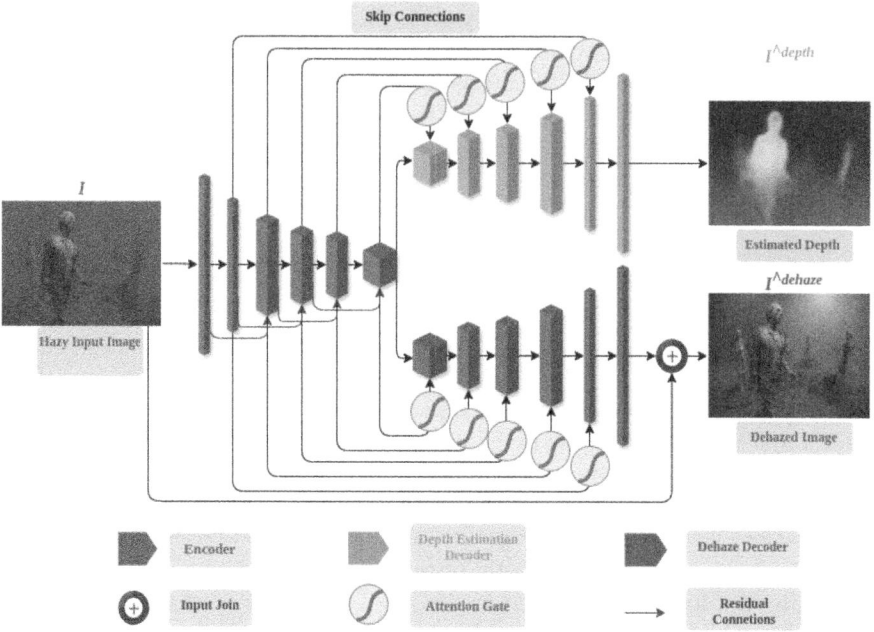

Fig. 1. Architecture of 2HDED:AttN for joint image de-hazing and depth estimation.

RGB images. 2HDED:AttN consists of three main parts, an encoder for feature extraction and two decoders for depth map estimation and RGB image dehazing.

Feature extraction is an important part of the deep learning models to extract descriptive features from input data at various levels. DenseNet-121 was used in [13] and provided promising results for open-air depth estimation. However, underwater imaging is a more challenging task that requires a higher capacity feature extraction. To this end, we utilize DenseNet-161 [8] as the feature extractor of the proposed 2HDED:AttN architecture to benefit from a deeper encoder and its higher capacity to learn complex features. DenseNet-161 has a higher top-1 accuracy on ImageNet and a higher top-5 accuracy, representing its outstanding capacity for image classification even in unclear environments (e.g., optically distorted underwater images).

In the next stage, two parallel decoders use the latent features to estimate the depth map and produce the dehazed RGB image. The architecture of the decoders is the same as the encoder network in reverse order.

The attention gate mechanism is introduced in [16] for U-Net architectures, and is used to focus the attention of the network on specific target structures and regions in the data and highlight only relevant information while training the model. Attention gates have been proven to improve the generalisability of the network. In the U-Net architecture, skip connections are used to combine the spatial information from the encoder with the decoder path. However, skip connections might bring redundant low-level features into the higher-level decoder layers.

Using the attention gate mechanism in the skip connections can actively reduce redundant features by suppressing their activation while bringing them to the decoder layers. The proposed 2HDED:AttN architecture uses the attention gate mechanism, increasing its generalizability and improving the ability of the decoder to utilize more relevant spatial information. The attention gates are shown on the skip connections in Fig. 1, while Fig. 2 illustrates the attention gate mechanism schematics.

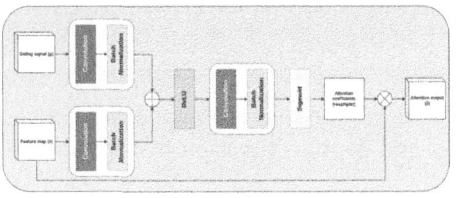

Fig. 2. Attention Gate Mechanism.

3.2 Network Training and Loss Function

Training of the 2HDED:AttN follows a supervised approach, requiring clear (i.e., haze-free) RGB images and their corresponding depth maps. This approach ensures effective learning and parameter refinement for improved performance in depth estimation and image dehazing.

The L_1 norm, known for its ability to estimate sparse solutions, is employed for depth estimation loss, L_1^{Depth}.

$$L_1^{Depth} = \frac{1}{n}\sum_{i=1}^{n}|\widehat{I}_i^{depth} - I_i^{depth}| \qquad (1)$$

where $\widehat{I}^{depth}$ is the estimated depth, I^{depth} the ground truth, i is the current pixel and n is the number of pixels.

Additionally, we augmented the L_1 norm with a smoothing regularization term aimed at eliminating low-amplitude structures within the depth map while enhancing the prominence of main edges, as suggested by [13].

$$L_{grad} = \frac{1}{n}\sum_{i}|\Delta_x R_i| + |\Delta_y R_i| \qquad (2)$$

where $R_i = \widehat{I}_i^{depth} - I_i^{depth}$ and Δ_x and Δ_y are the spatial derivatives with respect to the x-axis and y-axis. The total loss function for depth estimation is as follows:

$$L_{depth} = L_1^{Depth} + \mu L_{grad} \qquad (3)$$

where μ is a weighting coefficient, set to 0.001.

Mean Squared Error (MSE) is the most common loss function for image dehazing and quality enhancement [4,9,11]. MSE is the average squared difference between the predicted and target images, as shown in Eq. 4.

$$L_{dehaze} = \frac{1}{2}\sum_{i=1}^{n}(I^{gt} - \widehat{I}^{dehaze}) \qquad (4)$$

where I^{gt} is the ground truth image and $\hat{I}^{dehaze}$ is the estimated dehazed image. With the depth and dehazing losses defined as in Eqs. 3 and 4, we define the following total loss for 2HDED:AttN training:

$$L_{2HDED} = L_{depth} + \lambda L_{dehaze} \tag{5}$$

where λ is the weighting coefficient, which is used to balance between the dehazing and depth estimation tasks and set to 0.01.

4 Experimental Results

In this section, first, the datasets used for training and validation of the 2HDED:AttN architecture, are introduced. Later, comprehensive quantitative and qualitative analyses are carried out to showcase the performance of the proposed method.

4.1 Datasets and Experimental Setup

For the experimental evaluation, we use Underwater Salient Object Detection (USOD10K) [7], and Underwater Image Enhancement Benchmark (UIEB) [9] datasets.

USOD10K dataset is collected from the internet using different search engines such as Google, Bing, and Baidu (a similar approach to ImageNet and Microsoft COCO datasets), as well as some scenes from existing underwater image datasets. Five different methods are used in the US0D10K dataset to generate simulated depth maps, including DCP, Underwater Dark Channel Prior (UDCP), Underwater Dark Channel Prior Plus (UDCP+), Underwater Watershed Network (UWNet), and Deep Photometric Transformer (DPT). The USOD10K dataset comprises underwater RGB images paired with corresponding ground truth depth maps, all with a pixel resolution of 640 × 480. In total, the dataset contains 10,255 underwater images with 70 categories of salient objects in 12 different underwater scenes. The official train and test split from the USOD10K dataset with 9229 training and 1026 testing samples are used. In addition to the ground truth depth maps, hazy RGB images are required to train the 2HDED:AttN network with two output heads. The Foggy and Hazy Images Simulator (FoHIS) from [23] is used to add synthetic haze to the RGB images in the USOD10K dataset. The method takes into account the elevation of each pixel in the image, which allows for a more accurate haze simulation than previous methods that only consider depth. The FoHIS method first estimates the elevation of all the pixels in the image. Then, it calculates the opacity of each pixel based on its elevation and the extinction coefficient of the atmosphere. Finally, it combines the attenuated light using the Beer-Lambert law of attenuation from the object with the scattered light from the atmosphere to create the simulated hazy image.

$$I = I_{ex} + O_p * I_{al} \tag{6}$$

In the Eq. 6, the simulated hazy image I is generated by combining the attenuated light I_{ex} from the object and the scattered light from the atmosphere. The opacity of the pixel O_p is used to weigh the amount of scattered light I_{al} added to the image. The higher the opacity of a pixel, the more scattered light is added to the image. An example of the haze simulation can be seen in Fig. 3.

The USOD10K dataset is used for training the 2HDED:AttN network and evaluating the depth estimation and image dehazing results in comparison to the other underwater deep learning-based models. Additionally, the UIEB dataset is exclusively used for image dehazing evaluation. This involved training the model using the USOD10K dataset and testing on the UIEB dataset for image dehazing.

Without Haze   Synthetic Haze

Fig. 3. Example of synthetic haze generation using Foggy and Hazy Images Simulator (FoHIS) [23].

The UIEB dataset comprises 890 raw real images affected by haze and their corresponding reference haze-free images.

The network is implemented using the PyTorch framework. The entire training session takes approximately 9 h on an NVIDIA Quadro GV100 GPU with 32 GB memory. We trained 2HDED:AttN for 500 epochs with a batch size of 16 images resized to 640 × 480. We use Stochastic Gradient Descent (SGD) optimizer with an initial learning rate of 0.0002. The initial learning rate is reduced 10 times after the first 300 epochs, this allows for large weight changes at the beginning of the learning process and small changes towards the end of the learning process. Our network has $41M$ trainable parameters, whereas [2] has $32M$. It is worth mentioning that our network jointly estimates depth maps and dehazed images with two decoders while [2] solely estimates the depth maps. Also, our network is trained on the images with a size of 640 × 480 while [2] trained their network on images with a much smaller size of 320 × 240.

4.2 Qualitative and Quantitative Evaluation

In this section, qualitative and quantitative evaluations of the proposed architecture are carried out. The 2HDED:AttN model is trained on the USOD10K training dataset and the depth map and RGB image dehazing are compared with the other deep learning-based models. Three metrics, including Root Mean Squared Error (RMSE), Absolute Relative Error (Abs Rel), and Logarithmic Error (Log10) are used for depth estimation quality assessment. For the image dehazing task, we assess the quality of the dehazed images using the Peak Signal-to-Noise Ratio (PSNR) and Structural Similarity Index (SSIM).

2HDED:AttN model outputs the depth map and dehazed RGB image, simultaneously, while most of the other deep learning-based models are specifically trained for one of these tasks. Only Nazir et al. [14] is comparable to the 2HDED:AttN model, in this regard. Chen et al. [2] is used for comparison between the depth map estimation with the USOD10K testing dataset, and

Qin et al. [18] and Espinosa et al. [4] are used for image dehazing comparison with this dataset. Li et al. [9] is also used for image dehazing comparison with UIEB as the testing dataset.

The quantitative results are shown in Table 1. The comparison between the 2HDED:AttN and Nazir et al. [14] models shows that adding attention gate mechanisms and employing a deeper feature extractor as the encoder significantly improve the depth estimation and image dehazing quality. 2HDED:AttN results in lower RMSE (0.214 vs. 0.301), Abs Rel (0.658 vs. 0.895) and Log10 (0.15 vs. 0.18) for depth estimation, and higher PSNR (27.03 vs. 25.8) and SSIM (0.87 vs. 0.82) for RGB image dehazing. These improvements highlight the efficacy of 2HDED:AttN in capturing complex features in a more complicated environment. Additionally, depth estimation and image dehazing results are compared in Figs. 4 and 5, respectively. The qualitative analyses and visual comparison demonstrate that the 2HDED:AttN model generates superior depth maps with finer details and edges, and clearer and more visually appealing dehazed images.

The other deep learning-based models are trained specifically for one task, as a result, we compare the depth estimation and image dehazing separately. Regarding depth estimation, we benchmarked our results against Chen et al. [2], who achieved promising results by incorporating Laplacian Pyramid-Based Depth Residuals in their Encoder-Decoder network. It is noteworthy that [2] focuses solely on depth estimation without considering the image dehazing task. Chen et al. [2] obtains a lower RMSE of 0.121 using $32M$ parameters for only depth estimation, compared to 0.214 achieved by 2HDED:AttN using $41M$ parameters for multiple tasks. The qualitative results are depicted in Fig. 6.

For image dehazing, we selected a general dehazing method [18] and Espinosa et al. [4] which is specifically trained and tested on underwater images. Quan-

Table 1. Quantitative results using 2HDED:AttN and comparison with other deep learning models for depth estimation and image dehazing. For RMSE, Abs Rel, Log10 lower is better and for PSNR and SSIM higher is better.

Model	Depth Estimation			Dehazing	
	RMSE ↓	Abs Rel ↓	Log10 ↓	PSNR ↑	SSIM ↑
USOD10K Dataset					
2HDED:AttN	0.21	0.65	0.15	27.03	0.87
Nazir et al. [14]	0.30	0.89	0.18	25.80	0.82
Chen et al. [2]	0.12	0.54	0.13	–	–
Qin et al. [18]	–	–	–	26.91	0.85
Espinosa et al. [4]	–	–	–	28.20	0.89
UIEB Dataset					
2HDED:AttN	–	–	–	15.82	0.59
Li et al. [9]	–	–	–	19.11	0.79

Fig. 4. Depth map comparison between 2HDED:AttN and [14].

Fig. 5. Dehazed RGB image comparison between 2HDED:AttN and [14].

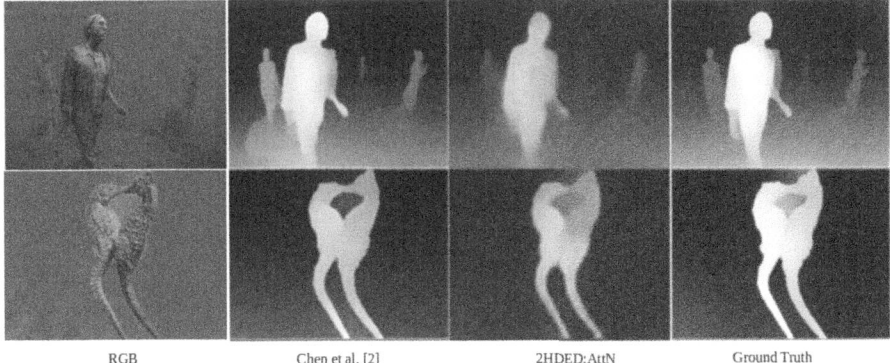

Fig. 6. Comparison of 2HDED:AttN with other deep learning-based underwater depth estimation model on USOD10K Dataset.

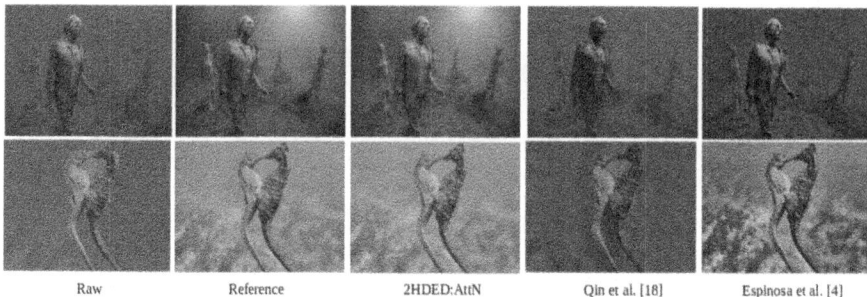

Fig. 7. Comparison of 2HDED:AttN with other deep learning-based underwater image dehazing based model on USOD10K Dataset.

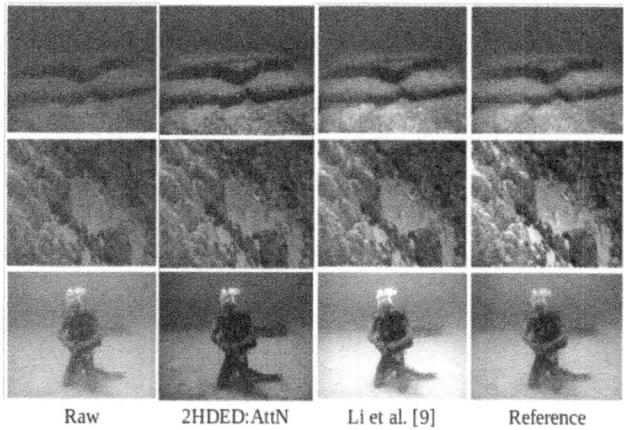

Fig. 8. Comparison of 2HDED:AttN with [9] for underwater image enhancement on UIEB Dataset.

titatively, the results obtained by [4] surpass those of 2HDED:AttN. However, 2HDED:AttN still generates better dehazing results than [18]. Similarly, Fig. 7 illustrates the comparable dehazed images of 2HDED:AttN to the [4,18].

Furthermore, the lower section of Table 1 and Fig. 8 compare the 2HDED:AttN model's dehazing output with [9], which focuses on underwater image enhancement using the real dataset UIEB. Despite being trained on a different dataset (USOD10K synthetic dataset), 2HDED:AttN obtains comparable results to that of [9] for dehazing UIEB images, demonstrating its generalizability.

5 Conclusion

This study proposes a novel deep learning architecture based on attention U-Net and DenseNet-161 networks, called 2HDED:AttN, for jointly estimating the depth map and dehazed RGB image from a single hazy underwater image. By parallelizing the depth estimation and image dehazing tasks, the complexity of the network is reduced, while preserving a superior or comparable performance to the state-of-the-art results. Quantitative and qualitative analyses demonstrated the superior performance of the 2HDED:AttN compared to the other architecture for jointly estimating the depth map and dehazed images and comparable results to the different models specifically trained for one task. These results show the accuracy and generalizability of the proposed 2HDED:AttN architecture. However, the 2HDED:AttN model can be improved in future studies by adding more advanced modules to the architecture and training with real data and diverse underwater environments.

Acknowledgments. This project has received funding from the Normandy region under the COSURIA project.

References

1. Ancuti, C., Ancuti, C.O., Haber, T., Bekaert, P.: Enhancing underwater images and videos by fusion. In: 2012 IEEE Conference on Computer Vision and Pattern Recognition, pp. 81–88. IEEE (2012)
2. Chen, W., Luo, X., Li, F., Wang, D.: Estimation of underwater monocular depth using Laplacian pyramid-based depth residuals. In: 2023 13th International Conference on Information Science and Technology (ICIST), pp. 40–47. IEEE (2023)
3. Drews, P.L., Nascimento, E.R., Botelho, S.S., Campos, M.F.M.: Underwater depth estimation and image restoration based on single images. IEEE Comput. Graphics Appl. **36**(2), 24–35 (2016)
4. Espinosa, A.R., McIntosh, D., Albu, A.B.: An efficient approach for underwater image improvement: Deblurring, dehazing, and color correction. In: Proceedings of the IEEE/CVF Winter Conference on Applications of Computer Vision, pp. 206–215 (2023)
5. Hambarde, P., Murala, S., Dhall, A.: UW-GAN: single-image depth estimation and image enhancement for underwater images. IEEE Trans. Instrum. Meas. **70**, 1–12 (2021)
6. He, K., Sun, J., Tang, X.: Single image haze removal using dark channel prior. IEEE Trans. Pattern Anal. Mach. Intell. **33**(12), 2341–2353 (2010)
7. Hong, L., Wang, X., Zhang, G., Zhao, M.: USOD10K: a new benchmark dataset for underwater salient object detection. IEEE Trans. Image Process. (2023)
8. Huang, G., Liu, Z., Van Der Maaten, L., Weinberger, K.Q.: Densely connected convolutional networks. In: 2017 IEEE Conference on Computer Vision and Pattern Recognition (CVPR), pp. 2261–2269 (2017). https://doi.org/10.1109/CVPR.2017.243
9. Li, C., et al.: An underwater image enhancement benchmark dataset and beyond. IEEE Trans. Image Process. **29**, 4376–4389 (2019)
10. Li, C., Guo, J., Guo, C.: Emerging from water: underwater image color correction based on weakly supervised color transfer. IEEE Signal Process. Lett. **25**(3), 323–327 (2018)

11. Li, J., Skinner, K.A., Eustice, R.M., Johnson-Roberson, M.: WaterGAN: unsupervised generative network to enable real-time color correction of monocular underwater images. IEEE Robot. Autom. Lett. **3**(1), 387–394 (2017)
12. Massot-Campos, M., Oliver-Codina, G.: Optical sensors and methods for underwater 3D reconstruction. Sensors **15**(12), 31525–31557 (2015)
13. Nazir, S., Vaquero, L., Mucientes, M., Brea, V.M., Coltuc, D.: 2HDED: net for joint depth estimation and image deblurring from a single out-of-focus image. In: 2022 IEEE International Conference on Image Processing (ICIP), pp. 2006–2010. IEEE (2022)
14. Nazir, S., Vaquero, L., Mucientes, M., Brea, V.M., Coltuc, D.: Depth estimation and image restoration by deep learning from defocused images. IEEE Trans. Comput. Imaging (2023)
15. Niu, Z., Hua, G., Gao, X., Tian, Q.: Context aware topic model for scene recognition. In: 2012 IEEE Conference on Computer Vision and Pattern Recognition, pp. 2743–2750. IEEE (2012)
16. Oktay, O., et al.: Attention U-Net: Learning where to look for the pancreas (2018). https://arxiv.org/abs/1804.03999
17. Pérez, J., Bryson, M., Williams, S.B., Sanz, P.J.: Recovering depth from still images for underwater dehazing using deep learning. Sensors **20**(16), 4580 (2020)
18. Qin, X., Wang, Z., Bai, Y., Xie, X., Jia, H.: FFA-Net: feature fusion attention network for single image dehazing. In: Proceedings of the AAAI Conference on Artificial Intelligence. vol. 34, pp. 11908–11915 (2020)
19. Shin, Y.S., Cho, Y., Pandey, G., Kim, A.: Estimation of ambient light and transmission map with common convolutional architecture. In: OCEANS 2016 MTS/IEEE Monterey, pp. 1–7. IEEE (2016)
20. Wang, Y., Zhang, J., Cao, Y., Wang, Z.: A deep CNN method for underwater image enhancement. In: 2017 IEEE International Conference on Image Processing (ICIP), pp. 1382–1386. IEEE (2017)
21. Ye, X., et al.: Deep joint depth estimation and color correction from monocular underwater images based on unsupervised adaptation networks. IEEE Trans. Circuits Syst. Video Technol. **30**(11), 3995–4008 (2019)
22. Zhang, H., Patel, V.M.: Densely connected pyramid dehazing network. In: Proceedings of the IEEE Conference on Computer Vision and Pattern Recognition, pp. 3194–3203 (2018)
23. Zhang, N., Zhang, L., Cheng, Z.: Towards simulating foggy and hazy images and evaluating their authenticity. In: Neural Information Processing: 24th International Conference, ICONIP 2017, Guangzhou, China, November 14-18, 2017, Proceedings, Part III 24, pp. 405–415. Springer (2017)
24. Zhang, S., Zhang, J., Fang, S., Cao, Y.: Underwater stereo image enhancement using a new physical model. In: 2014 IEEE International Conference on Image Processing (ICIP), pp. 5422–5426. IEEE (2014)
25. Zhao, Q., Xin, Z., Yu, Z., Zheng, B.: Unpaired underwater image synthesis with a disentangled representation for underwater depth map prediction. Sensors **21**(9), 3268 (2021)
26. Zhou, G., Li, C., Zhang, D., Liu, D., Zhou, X., Zhan, J.: Overview of underwater transmission characteristics of oceanic lidar. IEEE J. Sel. Top. Appl. Earth Observations Remote Sens. **14**, 8144–8159 (2021)

KD-AHOSVD: Neural Network Compression via Knowledge Distillation and Tensor Decomposition

Laura Meneghetti[1](✉) , Edoardo Bianchi[2] , Nicola Demo[1] , and Gianluigi Rozza[1]

[1] Mathematics Area, mathLab, SISSA, Trieste, Italy
{laura.meneghetti,nicola.demo,gianluigi.rozza}@sissa.it
[2] Free University of Bozen-Bolzano, Bolzano, Italy
edbianchi@unibz.it

Abstract. In the field of Deep Learning, the high number of parameters in models has become a significant concern within the scientific community due to the increased computational resources and memory required for training and inference. Addressing this issue, we propose a novel tensorized technique to compress network architectures. Our approach aims to significantly reduce the network's size and the number of parameters by integrating Averaged Higher Order Singular Value Decomposition with a novel Knowledge Distillation approach. Specifically, we replace certain layers of the original architecture with layers that perform linear projections onto a reduced space defined by our reduction technique. We conducted experiments on image classification tasks using multiple architectures and datasets. The evaluation focuses on final accuracy, model size, and parameter reduction, comparing our approach with both the original models and quantization, a widely used reduction method. The results underscore the effectiveness of our method in significantly reducing the number of parameters and the overall size of neural networks while maintaining high performance.

Keywords: Network Compression · Image Processing · Tensor Decomposition

1 Introduction

The growing adoption of embedded vision systems in industry has highlighted key limitations of standard Deep Neural Networks (DNNs), particularly their high parameter count, lengthy training times, and significant memory requirements [23,31]. To address these challenges, various network compression techniques have emerged, including pruning, quantization, low-rank decomposition, and Knowledge Distillation (KD) [22]. Building on these advances, we use Tensor Decomposition (TD) to compress DNNs for embedded applications, leveraging previous works [19,25] to enhance the reduction framework developed by Cui et al. [6] and Meneghetti et al. [20,21] for image classification tasks. Our proposed method combines Averaged Higher Order Singular Value Decomposition (AHOSVD), a variant of a commonly used tensor decomposition technique, with KD [12], selectively reducing the original model's layers while connecting them to a predictor block through a reduction layer. This reduction layer uses AHOSVD to project high-dimensional outputs to lower dimensions, retaining key parameters, while KD allows knowledge transfer from a larger model to the compressed one, preserving performance despite the model's reduced size.

The main contributions of this paper are summarized as follows:

- We introduce AHOSVD, an advanced version of HOSVD that uses an averaged Singular Value Decomposition (SVD) across image batches, significantly reducing computational complexity while maintaining accuracy.
- We propose a novel neural network compression pipeline that segments the original network, incorporates AHOSVD-based reduction layers, and adds fully connected layers for classification. The compressed network is fine-tuned with our KD approach to enhance performance and generalization.

Experiments on CIFAR-10, CIFAR-100, and STL-10 across various architectures demonstrate the effectiveness of our approach, which achieves higher compression rates than quantization techniques.

2 Background and Related Works

Various network compression techniques have been developed to address the limitations of Deep Neural Network architectures, such as network pruning [11, 16], parameter quantization [33], and tensor factorization [5,17], as well as KD [8,12]. The introduction of reduction layers through Model Order Reduction techniques like Active Subspaces (AS) and Proper Orthogonal Decomposition (POD) has also been suggested [6,20,21]. However, these methods often lose the tensorial structure of the data, reducing multidimensional outputs to vectors and mixing critical features.

Recently, Tensor Decompositions (TDs) have gained interest in preserving the multidimensional structure of data in image processing [14,25]. TDs decompose data tensors into factor matrices, mitigating the curse of dimensionality [18]. Notable examples include Canonical Polyadic Decomposition (CPD) [4,10] and Tucker Decomposition (TKD) [29,30], with the latter often used for compression and dimensionality reduction. TKD's higher-order generalization of PCA, known as Multilinear SVD (MLSVD) or Higher Order SVD (HOSVD) [1,7], and Tensor Networks like Tensor Trains (TTs) [3,24], are also significant.

3 Proposed Methodology

In this section, we detail our proposed methodology for compressing neural network architectures using a combination of AHOSVD and our novel KD. We validate our methodology using ResNet101 and VGG19 as backbones and three popular datasets: CIFAR-10, CIFAR-100 and STL-10.

3.1 Averaged Higher Order Singular Value Decomposition (AHOSVD)

In this section, we describe the mathematical foundations and operations behind the AHOSVD used in our compression methodology. AHOSVD extends the traditional HOSVD to provide a more computationally efficient approach for large-scale datasets.

Higher Order Singular Value Decomposition (HOSVD). HOSVD generalizes the matrix SVD to higher-order tensors. Given a tensor $\mathcal{A} \in \mathbb{R}^{I_1 \times I_2 \times \cdots \times I_N}$, the HOSVD decomposes it into a core tensor $\mathcal{S}$ and a set of orthogonal matrices $\mathbf{U}^{(n)} \in \mathbb{R}^{I_n \times I_n}$ for $n = 1, \ldots, N$:

$$\mathcal{A} = \mathcal{S} \times_1 \mathbf{U}^{(1)} \times_2 \mathbf{U}^{(2)} \cdots \times_N \mathbf{U}^{(N)}, \tag{1}$$

where $\times_n$ denotes the mode-n tensor-matrix product. The core tensor $\mathcal{S}$ captures the interactions between different modes, and the matrices $\mathbf{U}^{(n)}$ contain the orthonormal vectors of the mode-n fibers.

Averaged HOSVD (AHOSVD). The AHOSVD algorithm, outlined in Algorithm 1, addresses HOSVD's computational and memory demands by processing the tensor $\mathcal{A}$ in smaller batches along the first dimension. This batch-wise approach reduces computational load and memory usage by averaging projection matrices, allowing large tensors to be managed without storing them in full. The method approximates these matrices through a truncated singular value decomposition averaged over the batch, maintaining computational efficiency with minimal impact on accuracy.

Given a tensor $\mathbf{A}$, the AHOSVD technique first computes the mode-n unfolding $\mathbf{A}_{(n)}$ for each mode n. The singular value decomposition (SVD) of $\mathbf{A}_{(n)}$ is:

$$\mathbf{A}_{(n)} = \mathbf{U}^{(n)} \mathbf{\Sigma}^{(n)} \mathbf{V}^{(n)T}. \tag{2}$$

Instead of directly using $\mathbf{U}^{(n)}$, AHOSVD uses an incremental average of this $\mathbf{U}^{(n)}$ returned at different steps of the procedure, i.e. while considering different batches of images:

$$\mathbf{U}^{(n)}_{\text{avg}} = \frac{n_{\text{avg}}}{n_{\text{avg}} + 1} \mathbf{U}^{(n)}_{\text{old}} + \frac{1}{n_{\text{avg}} + 1} \mathbf{U}^{(n)}_{\text{new}}, \tag{3}$$

where n_{avg} is the number of elements the current average is taken over. This averaged matrix is then used to construct the projection matrices used for reduction by keeping only the first R columns, corresponding to the highest singular values of the averaged $\mathbf{\Sigma}^{(n)}$. An eigenvalue analysis can be performed in each tensor direction to determine a suitable value for R, as done in [6,21].

By using AHOSVD, we can efficiently reduce the dimensionality of the tensor data while preserving the essential features needed for accurate neural network predictions.

Algorithm 1: Averaged Higher Order Singular Value Decomposition (AHOSVD)

Input: $(N+1)$-dimensional tensor $A \in \mathbb{R}^{I_0 \times I_1 \times \cdots \times I_N}$, batch size $K \in \mathbb{N}$
Output: matrices $U^{(1)}, \ldots, U^{(N)}$

1. initialize the zero matrices $U^{(1)} \in \mathbb{R}^{I_1 \times I_1}, \ldots, U^{(N)} \in \mathbb{R}^{I_N \times I_N}$;
2. assert $I_0 = 0 \mod K$;
3. define $q = I_0/K$;
4. **for** $j = 0, 1, \ldots, q-1$ **do**
5. $\quad$ define $A_{[j]} = A[jK+1 : jK+K, :, \ldots, :] \in \mathbb{R}^{K \times I_1 \times \cdots \times I_N}$;
6. $\quad U^{(0)}_{[j]}, U^{(1)}_{[j]}, \ldots, U^{(N)}_{[j]}, S_{[j]} = \text{HOSVD}(A_{[j]})$;
7. $\quad U^{(1)}, \ldots, U^{(N)} = \text{UpdateAvg}([U^{(1)}, \ldots, U^{(N)}], [U^{(1)}_{[j]}, U^{(2)}_{[j]}, \ldots, U^{(N)}_{[j]}], j)$;
8. **return** $U^{(1)}, U^{(2)}, \ldots, U^{(N)}$;

3.2 Reduction Pipeline

Starting from the reduction methods introduced by Meneghetti et al., [20,21] we provide a general framework applicable to Convolutional Neural Networks (CNNs). Our approach leverages the AHOSVD to reduce the dimensionality of the network while integrating a novel KD paradigm to fine-tune the compressed model for downstream tasks.

Algorithm 2: Pseudo-code for model reduction.

Input:
- Original (pre-trained) network $\mathcal{ANN}$
- Fine-tuning dataset $\mathcal{D}_{\text{train}} = \{\mathbf{x}^{(0),j}, \mathbf{y}^j\}_{j=1}^{N_{\text{train}}}$
- Cut-off index $\ell \in \mathbb{N}$
- Reduced dimension R

1. $\mathcal{ANN}^\ell_{\text{pre}}, \mathcal{ANN}^\ell_{\text{post}} = \text{Splitting}(\mathcal{ANN}, \ell)$;
2. $\mathbf{x}^{(\ell),j} = \mathcal{ANN}^\ell_{\text{pre}}(\mathbf{x}^{(0),j})$ for $j = 1, \ldots, N_{\text{train}}$;
3. $X = [\mathbf{x}^{(\ell),1}, \ldots, \mathbf{x}^{(\ell),N_{\text{train}}}]$;
4. $Z = \text{Reduction}(X, R)$;
5. $\mathcal{ANN}^{\text{red}} = \text{Predictor}(Z, \text{additional inputs})$;
6. Fine-tuning (KD) $\mathcal{ANN}^{\text{red}}$ on $\mathcal{D}_{\text{train}}$;

Output: Reduced fine-tuned network $\mathcal{ANN}^{\text{red}}$

Algorithm 2 summarizes the fundamental steps to construct a reduced version of a general ANN. The process begins with splitting the original pre-trained network into two parts based on a specified cut-off index, referring to a specific ANN layer. This results in a pre-model, which handles the initial feature extraction, and a post-model. The pre-model is retained for further processing, while the post-model is discarded.

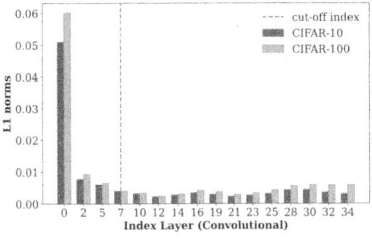
(a) VGG19 with CIFAR datasets

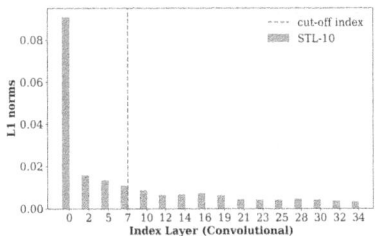
(b) VGG19 with STL-10

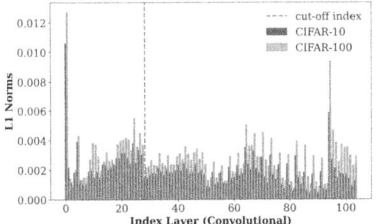
(c) ResNet101 with CIFAR datasets

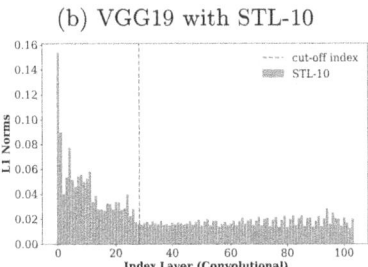
(d) ResNet101 with STL-10

Fig. 1. Structural analysis of deep neural networks by L1-norm. All experiments are conducted on VGG19 (Fig. 1a and Fig. 1b) and on ResNet101 (Fig. 1c and Fig. 1d) using CIFAR-10, CIFAR-100, and STL-10 datasets. Layer indices refer to a list where only convolutional and linear layers are considered as possible cut-off layers. Dotted lines indicate the chosen cut-off indices.

The cut-off index selection is guided by an L1-norm analysis, where we evaluate the L1-norm, normalized by parameter count, across each trainable layer. The cut-off is chosen at the layer where the L1-norm stabilizes, with subsequent values remaining comparable or lower, balancing storage efficiency with model performance. This approach, common in pruning methods [15], identified cut-off values of 7 for VGG19 and 28 for ResNet101, considering only convolutional and linear layers (Fig. 1). However, it is also possible to select larger cut-off indices to adjust the trade-off between compression and accuracy, allowing flexibility based on specific application requirements.

Next, the input data is processed through the pre-model to generate high-dimensional features, which are reduced to lower dimensions using AHOSVD, minimizing computational load while retaining essential information. Following dimensionality reduction, a classifier is added to the compressed network, comprising Fully Connected (FC) layers for classification.

Finally, the reduced network undergoes fine-tuning using a novel KD approach to align its performance with that of the original model. Unlike standard KD, which transfers information via teacher logits, our method aligns classifier inputs for the teacher and student using a regressor to bridge dimension differences. This approach aligns with methods in [28,32], which leverage intermediate representations to enhance student performance. Figure 2 illustrates the reduction process, transitioning from the pre-trained network to the fine-tuned, reduced model.

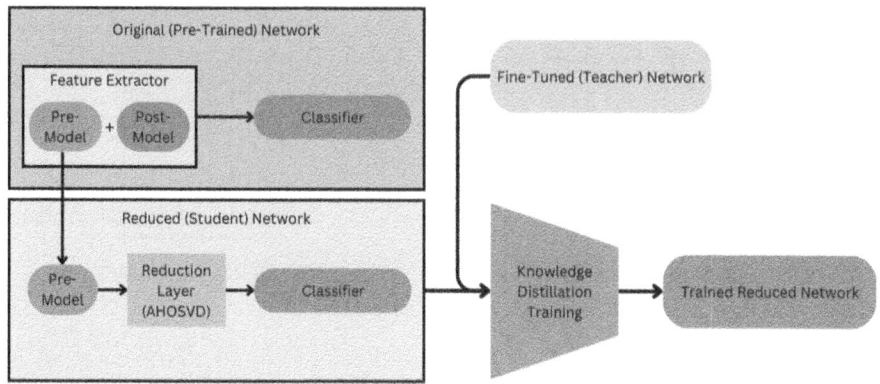

Fig. 2. Overview of the KD-AHOSVD compression approach. The original pre-trained network (teacher) is split into pre-model and post-model components at the cut-off layer. The pre-model's output is processed through a reduction layer (AHOSVD) and combined with a classifier to create the reduced (student) network. The student network is then fine-tuned using KD training from the fine-tuned teacher network, resulting in the final compressed model.

Technically, given a network $\mathcal{ANN} : \mathbb{R}^{n_0} \to \mathbb{R}^{n_L}$ with L hidden layers and a cut-off index ℓ, the model is split into two parts: the *pre-model* and the *post-model*. Describing $\mathcal{ANN}$ as the composition of L functions $f_j : \mathbb{R}^{n_{j-1}} \to \mathbb{R}^{n_j}$, for $j = 1, \ldots, L$, the pre- and post-models are defined by:

$$\mathcal{ANN}^\ell_{\text{pre}} = f_\ell \circ f_{\ell-1} \circ \cdots \circ f_1, \tag{4}$$
$$\mathcal{ANN}^\ell_{\text{post}} = f_L \circ f_{L-1} \circ \cdots \circ f_{\ell+1}.$$

Thus, the full model can be described as:

$$\mathcal{ANN}(\mathbf{x}^{(0)}) \equiv \mathcal{ANN}^\ell_{\text{post}}(\mathcal{ANN}^\ell_{\text{pre}}(\mathbf{x}^{(0)})), \tag{5}$$

where $\mathbf{x}^{(0)} \in \mathbb{R}^{n_0}$ is an input image from the train dataset $\mathcal{D}_{\text{train}} = \{\mathbf{x}^{(0),j}, \mathbf{y}^j\}_{j=1}^{N_{\text{train}}}$. For the pre-model outputs $\mathbf{x}^{(\ell)}$, these tensors belong to a high-dimensional space, necessitating a reduction map. Using the proposed AHOSVD, we define X as the input tensor containing all pre-model outputs. Through the eigenvalue analysis mentioned in Sect. 3.1, a suitable value for reduction can be determined. Given $R = (R_1, R_2, R_3)$ as the reduced dimensions, the compressed solution tensor $Z \in \mathbb{R}^{N_{\text{train}} \times R_1 \times R_2 \times R_3}$ is obtained by:

$$Z = X \times_1 U^{(1)}_{R_1} \times_2 U^{(2)}_{R_2} \times_3 U^{(3)}_{R_3}, \tag{6}$$

where $U^{(i)}_{R_i}$ are computed using AHOSVD.

After the reduction, the resulting compressed network $\mathcal{ANN}^{\text{red}}$ is fine-tuned on the downstream task dataset $\mathcal{D}_{\text{train}}$ employing the aforementioned KD approach. In this setup, the teacher is the non-reduced version of the original network fine-tuned on the same downstream task dataset $\mathcal{D}_{\text{train}}$, while the student is the reduced network $\mathcal{ANN}^{\text{red}}$.

In the novel KD process introduced, the student network's reduction layer output is aligned with the feature extractor output of the teacher network, minimizing discrepancies before classification. A convolutional regression layer compensates for dimensionality differences, and a combined loss function is used for training: Cross-Entropy for classification accuracy and Mean Squared Error (MSE) to match teacher-student classifier inputs. This approach enables the student network to retain essential knowledge from the teacher while maintaining a compact size.

4 Experiments and Results

We evaluate the KD-AHOSVD reduction pipeline on VGG19 and ResNet101 using CIFAR-10, CIFAR-100, and STL-10, comparing it to post-training quantization (Int2, Int4, Int8). Quantization is chosen as a baseline due to its practicality in resource-limited deployments, broad impact on key metrics like model size and accuracy, and alignment with modern hardware optimized for integer operations [9].

4.1 Datasets

We evaluate our approach on three popular datasets:

CIFAR-10: Contains 60,000 32×32 color images in 10 classes, with 50,000 for training and 10,000 for testing.

CIFAR-100: Similar to CIFAR-10 but with 100 classes, each containing 600 images (500 for training, 100 for testing).

STL-10: Consists of 96×96 color images in 10 classes, with 5,000 labeled training images, 8,000 test images, and 100,000 unlabeled images for unsupervised or semi-supervised learning.

4.2 Implementation Details

All experiments are implemented in PyTorch [26], with input data resized and normalized. We evaluate the proposed AHOSVD-based compression on two architectures, ResNet101 and VGG19. Each model is fine-tuned using our novel KD approach, with the original, non-reduced model fine-tuned on each benchmark dataset serving as the teacher. All evaluations are performed on a single NVIDIA A100 GPU with 40GB of memory.

Teacher Networks Training Pipeline. The non-reduced teacher networks enhance the performance of the reduced models. These teacher networks are fine-tuned on CIFAR-10, CIFAR-100, and STL-10. We apply quantization to the teacher models using Hugging Face's Optimum Quanto library [13], quantizing weights to Int2, Int4, and Int8, while keeping activations at Int8.

Teacher models are trained with Stochastic Gradient Descent (SGD) [27] using a learning rate of 0.01, momentum of 0.9, and weight decay of 10^{-4}, with a cosine annealing scheduler to reduce the learning rate. Training runs with a batch size of 24 for 120 epochs on CIFAR-10 and 200 epochs on CIFAR-100 and STL-10.

Student Networks Training Pipeline. The reduced student networks, initialized with ImageNet pre-trained weights, are fine-tuned on CIFAR-10, CIFAR-100, and STL-10 using our KD paradigm. Based on Fig. 1, the cut-off indices are set to 7 for VGG19 and 28 for ResNet101, with a reduced dimension $R = (50, 3, 3)$. The final classifier includes one hidden FC layer with 50 neurons for STL-10 and CIFAR-10, and 250 neurons for CIFAR-100.

Student networks are trained with an SGD optimizer at a learning rate of 10^{-4}, with a batch size of 24 for 120 epochs on CIFAR-10 and 200 epochs on CIFAR-100 and STL-10.

4.3 Evaluation Criteria

To evaluate our reduction approach, we compare KD-AHOSVD with multiple quantization approaches.

We assess performance using the following metrics:

- *Number of Parameters*: Total count of model learnable parameters.
- *Model Size (MB)*: Storage size of the model.
- *Top-1 Accuracy*: Percentage of test samples for which the model's top prediction is correct.
- *Top-5 Accuracy*: Percentage of test samples where the correct label is within the model's top 5 predictions.

4.4 Results

We present the performance of KD-AHOSVD compared to standard quantization techniques on VGG19 and ResNet101 across CIFAR-10, CIFAR-100, and STL-10 datasets, focusing on parameters, model size, and accuracy. Results are reported in Table 1.

Compression and Storage. KD-AHOSVD achieved significant model size reduction, outperforming quantization. For CIFAR-10, VGG19 was reduced from 558.44 MB to 7.03 MB (a 99% reduction), and ResNet101 from 170.5 MB to 16.46 MB (a 96% reduction). In comparison, quantization compressed VGG19 to around 140 MB and ResNet101 to 44 MB. Similar patterns were observed for CIFAR-100 and STL-10, where KD-AHOSVD consistently reduced model sizes by over 95%.

Accuracy. KD-AHOSVD maintained competitive accuracy with minimal reduction. For CIFAR-10, VGG19 achieved 89.66% top-1 accuracy (compared to 94.35% uncompressed), outperforming Int2/Int8 quantization (80.16%). Across CIFAR-100 and STL-10, KD-AHOSVD showed similarly minor accuracy drops relative to substantial compression gains, establishing it as a highly efficient method.

Table 1. Comparison of model performance metrics for VGG19 and ResNet101 on CIFAR-10, CIFAR-100, and STL-10 datasets using different compression approaches. The notation "Quant. (Wx/Ax)" indicates that weights are quantized to Wx bits and activations are quantized to Ax bits.

Dataset	Model	Compression	# Params	Size (MB)	Accuracy (%)	
					Top-1	Top-5
CIFAR-10	VGG19	–	139,611,210	558.44	94.35	99.88
		Quant. (Int2/Int8)	139,611,210	144.01	89.16	99.54
		Quant. (Int4/Int8)	139,611,210	144.01	94.26	99.87
		Quant. (Int8/Int8)	139,611,210	139.70	94.34	99.88
		KD-AHOSVD	1,757,854	7.03	89.66	99.36
	ResNet101	–	42,520,650	170.50	96.43	99.92
		Quant. (Int2/Int8)	42,520,650	44.58	67.99	94.09
		Quant. (Int4/Int8)	42,520,650	44.58	95.92	99.91
		Quant. (Int8/Int8)	42,520,650	43.47	95.70	99.90
		KD-AHOSVD	4,096,674	16.46	93.32	99.80
CIFAR-100	VGG19	–	139,979,940	559.92	76.86	94.31
		Quant. (Int2/Int8)	139,979,940	144.39	57.62	81.40
		Quant. (Int4/Int8)	139,979,940	144.39	76.73	94.19
		Quant. (Int8/Int8)	139,979,940	140.07	76.87	94.31
		KD-AHOSVD	3,542,616	14.17	65.72	89.28
	ResNet101	–	42,705,060	171.24	82.08	95.85
		Quant. (Int2/Int8)	42,705,060	44.77	11.48	29.59
		Quant. (Int4/Int8)	42,705,060	44.77	80.00	94.84
		Quant. (Int8/Int8)	42,705,060	43.65	80.64	95.37
		KD-AHOSVD	4,225,294	16.97	70.89	92.07
STL-10	VGG19	–	139,611,210	558.44	91.25	99.61
		Quant. (Int2/Int8)	139,611,210	144.01	85.88	99.19
		Quant. (Int4/Int8)	139,611,210	144.01	91.23	99.61
		Quant. (Int8/Int8)	139,611,210	139.70	91.23	99.60
		KD-AHOSVD	1,757,854	7.03	86.30	99.25
	ResNet101	–	42,520,650	170.50	95.01	99.92
		Quant. (Int2/Int8)	42,520,650	44.58	9.32	50.48
		Quant. (Int4/Int8)	42,520,650	44.58	90.01	99.56
		Quant. (Int8/Int8)	42,520,650	43.47	94.82	99.92
		KD-AHOSVD	4,096,674	16.46	90.69	99.54

5 Limitations

While KD-AHOSVD offers substantial benefits, it has limitations: high computational demands, untested performance on tasks beyond image classification, and added complexity from hyperparameter tuning. Additionally, compression and distillation may impact interpretability, suggesting areas for further research to enhance its applicability across domains.

6 Real-World Applications

The KD-AHOSVD compression method is well-suited for resource-constrained environments where computational power and memory are limited. In embedded vision systems (e.g., surveillance, drones, smart devices) and edge computing (e.g., AR, VR, real-time image processing), KD-AHOSVD enables efficient deployment of compact models. Industrial action recognition could also benefit, using efficient models to enhance safety and quality in resource-limited settings [2].

7 Discussion and Conclusion

The results in this paper demonstrate the effectiveness of KD-AHOSVD for neural network compression. By combining tensor decomposition with knowledge distillation, our approach achieves significant reductions in model size and parameters while maintaining high classification accuracy on CIFAR-10, CIFAR-100, and STL-10. KD-AHOSVD surpasses traditional quantization methods in compression, balancing efficiency and accuracy, making it suitable for resource-constrained environments like mobile and embedded systems.

In conclusion, KD-AHOSVD is a novel compression approach that integrates knowledge distillation with tensor decomposition, achieving notable model size and parameter reductions with minimal accuracy impact. This makes it a strong candidate for deploying deep learning models in environments with limited computational resources. Future work may explore its application to additional tasks and optimize its inference efficiency.

Acknowledgments. This work has been conducted within the research activities of the consortium iNEST (Interconnected North-East Innovation Ecosystem), (PNRR)-Missione 4 Componente 2, Investimento 1.5-D.D. 1058 23/06/2022, ECS00000043 supported by the European Union's NextGenerationEU program.

Disclosure of Interests. The authors have no competing interests to declare that are relevant to the content of this article.

References

1. Ahmadi-Asl, S., et al.: Randomized algorithms for computation of tucker decomposition and Higher Order SVD (HOSVD). IEEE Access **9**, 28684–28706 (2021). https://doi.org/10.1109/ACCESS.2021.3058103
2. Bianchi, E., Lanz, O.: Egocentric video-based human action recognition in industrial environments. In: Concli, F., Maccioni, L., Vidoni, R., Matt, D.T. (eds.) Latest Advancements in Mechanical Engineering, pp. 257–267. Springer Nature Switzerland, Cham (2024). https://doi.org/10.1007/978-3-031-70465-9_25
3. Brandoni, D., Simoncini, V.: Tensor-train decomposition for image recognition. Calcolo **57**, 1–24 (2020). https://doi.org/10.1007/s10092-020-0358-8
4. Carroll, J.D., Chang, J.J.: Analysis of individual differences in multidimensional scaling via an N-way generalization of "Eckart-Young" decomposition. Psychometrika **35**(3), 283–319 (1970). https://doi.org/10.1007/BF02310791
5. Cichocki, A., Lee, N., Oseledets, I., Phan, A.H., Zhao, Q., Mandic, D.P., et al.: Tensor networks for dimensionality reduction and large-scale optimization: Part 1 low-rank tensor decompositions. Foundations and Trends® in Machine Learning **9**(4-5), 249–429 (2016). https://doi.org/10.1561/2200000059
6. Cui, C., Zhang, K., Daulbaev, T., Gusak, J., Oseledets, I., Zhang, Z.: Active subspace of neural networks: structural analysis and universal attacks. SIAM J. Math. Data Sci. **2**(4), 1096–1122 (2020). https://doi.org/10.1137/19M1296070
7. De Lathauwer, L., De Moor, B., Vandewalle, J.: A multilinear singular value decomposition. SIAM J. Matrix Anal. Appl. **21**(4), 1253–1278 (2000). https://doi.org/10.1137/S0895479896305696
8. Gou, J., Yu, B., Maybank, S.J., Tao, D.: Knowledge distillation: a survey. Int. J. Comput. Vision **129**(6), 1789–1819 (2021). https://doi.org/10.1007/s11263-021-01453-z
9. Han, S., Mao, H., Dally, W.J.: Deep compression: compressing deep neural network with pruning, trained quantization and huffman coding. In: Bengio, Y., LeCun, Y. (eds.) 4th International Conference on Learning Representations, ICLR 2016, Conference Track Proceedings (2016)
10. Harshman, R.A.: Foundations of the PARAFAC procedure: models and conditions for an "explanatory" multi-modal factor analysis. UCLA Working Pap. Phonetics **16**, 1–84 (1970)
11. He, Y., Zhang, X., Sun, J.: Channel pruning for accelerating very deep neural networks. In: 2017 IEEE International Conference on Computer Vision (ICCV), pp. 1398–1406 (2017). https://doi.org/10.1109/ICCV.2017.155
12. Hinton, G., Vinyals, O., Dean, J.: Distilling the Knowledge in a Neural Network. In: NIPS Deep Learning and Representation Learning Workshop (2015)
13. Hugging Face: Optimum quanto: A PyTorch quantization backend for optimum. https://github.com/huggingface/optimum-quanto
14. Kolda, T.G., Bader, B.W.: Tensor decompositions and applications. SIAM Rev. **51**(3), 455–500 (2009). https://doi.org/10.1137/07070111X
15. Kumar, A., Shaikh, A.M., Li, Y., Bilal, H., Yin, B.: Pruning filters with l1-norm and capped l1-norm for CNN compression. Appl. Intell. **51**, 1152–1160 (2021). https://doi.org/10.1007/s10489-020-01894-y
16. Li, Y., Adamczewski, K., Li, W., Gu, S., Timofte, R., Van Gool, L.: Revisiting random channel pruning for neural network compression. In: 2022 IEEE/CVF Conference on Computer Vision and Pattern Recognition (CVPR), pp. 191–201 (2022). https://doi.org/10.1109/CVPR52688.2022.00029

17. Li, Y., Gu, S., Van Gool, L., Timofte, R.: Learning filter basis for convolutional neural network compression. In: 2019 IEEE/CVF International Conference on Computer Vision (ICCV), pp. 5622–5631 (2019). https://doi.org/10.1109/ICCV.2019.00572
18. Liu, X., Parhi, K.K.: Tensor decomposition for model reduction in neural networks: a review [feature]. IEEE Circuits Syst. Mag. **23**(2), 8–28 (2023). https://doi.org/10.1109/MCAS.2023.3267921
19. Liu, Y.: Tensors for data processing: theory, methods, and applications. Acad. Press (2021). https://doi.org/10.1016/C2020-0-01790-1
20. Meneghetti, L., Demo, N., Rozza, G.: A proper orthogonal decomposition approach for parameters reduction of single shot detector networks. In: 2022 IEEE International Conference on Image Processing (ICIP), pp. 2206–2210 (2022). https://doi.org/10.1109/ICIP46576.2022.9897513
21. Meneghetti, L., Demo, N., Rozza, G.: A dimensionality reduction approach for convolutional neural networks. Appl. Intell. 1–16 (2023). https://doi.org/10.1007/s10489-023-04730-1
22. Menghani, G.: Efficient deep learning: a survey on making deep learning models smaller, faster, and better. ACM Comput. Surv. **55**(12), 1–37 (2023). https://doi.org/10.1145/3578938
23. Messaoud, S., Bouaafia, S., Maraoui, A., Ammari, A.C., Khriji, L., Machhout, M.: Deep convolutional neural networks-based hardware–software on-chip system for computer vision application. Comput. Electr. Eng. **98**, 107671 (2022). https://doi.org/10.1016/j.compeleceng.2021.107671
24. Oseledets, I.V.: Tensor-train decomposition. SIAM J. Sci. Comput. **33**(5), 2295–2317 (2011). https://doi.org/10.1137/090752286
25. Panagakis, Y., et al.: Tensor methods in computer vision and deep learning. Proc. IEEE **109**(5), 863–890 (2021). https://doi.org/10.1109/JPROC.2021.3074329
26. Paszke, A., et al.: PyTorch: An Imperative Style, High-Performance Deep Learning Library. Curran Associates Inc., Red Hook, NY, USA (2019)
27. Robbins, H.E.: A stochastic approximation method. Ann. Math. Stat. **22**, 400–407 (1951)
28. Romero, A., Ballas, N., Kahou, S.E., Chassang, A., Gatta, C., Bengio, Y.: FitNets: hints for thin deep nets. In: Bengio, Y., LeCun, Y. (eds.) 3rd International Conference on Learning Representations, ICLR (2015)
29. Tucker, L.R.: The extension of factor analysis to three-dimensional matrices. In: Gulliksen, H., Frederiksen, N. (eds.) Contributions to Mathematical Psychology., pp. 110–127. Holt, Rinehart and Winston, New York (1964)
30. Tucker, L.R.: Some mathematical notes on three-mode factor analysis. Psychometrika **31**(3), 279–311 (1966). https://doi.org/10.1007/BF02289464
31. Udendhran, R., Balamurugan, M., Suresh, A., Varatharajan, R.: Enhancing image processing architecture using deep learning for embedded vision systems. Microprocess. Microsyst. **76**, 103094 (2020). https://doi.org/10.1016/j.micpro.2020.103094
32. Yang, J., Martinez, B., Bulat, A., Tzimiropoulos, G.: Knowledge distillation via softmax regression representation learning. In: International Conference on Learning Representations (2021)
33. Yang, J., et al.: Quantization networks. In: 2019 IEEE/CVF Conference on Computer Vision and Pattern Recognition (CVPR), pp. 7300–7308 (2019). https://doi.org/10.1109/CVPR.2019.00748

Analysis of Emerging Techniques for Signal Processing Applications

Novel Scheduling and Shifter Networks for 5G LDPC Decoders

Nikos Papageorgiou(✉) and Vassilis Paliouras

Department of Electrical and Computer Engineering, University of Patras, Patras, Greece
up1072843@ac.upatras.gr

Abstract. Low Density Parity Check (LDPC) codes under iterative decoding have shown remarkable error correction capabilities, with moderate complexity requirements. Initially, in this paper a check based on a part of the parity check matrix (core part check) is presented. The proposed check is amenable for hardware implementation and allows the termination of the decoding procedure at a sub-iteration level, i.e., within an iteration. In this way the number of clock cycles required reduced by 150 for the semi-parallel architecture and for 5G NR codes. Simultaneously the Block Error Rate (BLER) remains the same while hardware becomes simpler. In addition, this paper introduces a novel scheduling scheme combined with the core part check and a syndrome-select logic. Experimental results are offered, assuming the parallel architecture, which show that the proposed rescheduling reduces the average number of required clock cycles per decoded word. Simultaneously improves the BLER utilizing exactly the same hardware. Specifically, targeting 5G NR LDPC codes, gains of 35% are achieved, for both the OMS and NMS algorithms. Furthermore, an innovative approach for the merge network of the reconfigurable barrel shifter is proposed. Finally, an ASIC implementation for semi-parallel architecture and the energy consumption gained with the proposed algorithm are presented.

Keywords: LDPC · 5G NR · Rescheduling · Barrel Shifter

1 Introduction

Low-density parity-check (LDPC) codes were first introduced by Gallager in 1962 [8]. Through initially overlooked, despite their strong error correcting capabilities, they gained attention in 1995 when MacKay and Neal showed they could approach Shannon's theoretical limit [16]. LDPC codes in 5G NR are irregular Quasi-Cyclic (QC) and described by two base graphs (BG1, BG2). A binary parity-check matrix is derived from a BG of size $M \times N$, and expanded to $(M \times Z_c) \times (N \times Z_c)$, where Z_c is called lifting size and ranges from 1 to 384 [1]. Each element of BG is substituted by either a diagonal matrix of size $Z_c \times Z_c$, shifted cyclically, or a zero matrix of the same size, as specified by the standard [1].

© The Author(s), under exclusive license to Springer Nature Switzerland AG 2025
J. Lorandel and A. Kamaleldin (Eds.): DASIP 2025, LNCS 15569, pp. 95–107, 2025.
https://doi.org/10.1007/978-3-031-87897-8_8

Iterative LDPC decoding relies on Belief Propagation (BP) and variations of it. To mitigate the complexity of BP in computing check-to-variable ($C2V$) messages, Fossorier et al. introduced the Min Sum algorithm (MS) [7]. Although MS reduces hardware requirements, it affects the block error rate (BLER). To address this issue, Chen and Fossorier proposed Offset MS (OMS) and Normalized MS (NMS) where a threshold (b) and a normalized factor (a) are introduced respectively on calculation of $C2V$ message [4]. Other MS variations further reduce hardware costs by simplifying the minimum calculations, such as using the first minimum to approximate the second, as imposed by Petrović et al. [18]. Other variations improve the achievable BLER [6,21], where the normalized and offset factors are not predefined constants but vary during decoding.

In LDPC iterative decoding, each iteration's output is examined, applying $x \cdot H^T = s$, where x is the decoded word estimation, H^T is the transpose of the parity-check matrix H, and s is the syndromes vector, to determine if it corresponds to a valid codeword. If not, the process continues until either a codeword is found, $s=0$, or a maximum number of iterations is reached.

A crucial aspect of decoding is early termination, which reduces both power consumption and the number of required clock cycles. Several techniques for early termination have been examined, with hard decision aided (HDA) [19] being the most prominent. In HDA decoding stops if two consecutive iterations produce the same output. Li et al. refined HDA by performing checks at sub-iteration level, comparing the signs of Log-Likelihood Ratios (LLR_s) across two layers [12]. If the signs match, a counter is incremented and if it exceeds a threshold, the decoding halts. Furthermore, Kim and Kim combine sign LLR preservation with $x \cdot H^T = s$ check, but for the core part of H [11]. In order to obtain faster convergence and reduce the number of iterations, various scheduling ideas for check node processing have been proposed. Casado et al. introduced the Residual Belief Propagation (RBP) algorithm [3], an Informed Dynamic Scheduling (IDS), where check nodes with the largest residuals are processed first. Kim et al. followed a similar approach, but residual refers to $V2C$ messages [10].

In this work, we present a new scheduling scheme for the check-node processing and variable-node updating, tailored to 5G codes. It utilizes layered decoding scheduling [9], where each base graph row is treated as a layer, and involves the core part of the parity check matrix. We compare the block error rate and the required number of necessary iterations for decoding for the basic whole H approach, the core part approach and the proposed idea, for the OMS and NMS algorithms. Furthermore, we propose a modified reconfigurable barrel shifter with a different merge network, able for multi-frame processing. Finally, we introduce a semi-parallel architecture that will serve as the model for our simulations.

The remainder of the paper is organized as follows: Sect. 2 analyzes the core part logic for early termination and presents the decoding performance and hardware requirements. Section 3 introduces the proposed approach which is evaluated in Sect. 4. Finally, Sect. 5 presents the novel barrel shifter and conclusions are discussed in Sect. 6.

Table 1. Number of check nodes (BG1-BG2) connected with a specific number of variable nodes and needed xor gates

Number of Connections	XOR Gates	Number of Check Nodes	
		Full H	Core Part
19	37	4–0	4–0
10	19	1–2	0–2
9	17	2–0	0–0
8	15	2–2	0–2
7	13	5–0	0–0
6	11	8–3	0–0
5	9	18–9	0–0
4	7	5–20	0–0
3	5	1–6	0–0

2 Core Part Approach

2.1 Early Termination Approximations

Two techniques for early termination are analyzed.

- Soft Decision-Full H
  For this criteria, after computing the bits of the codeword $x_n = \frac{1-\text{sgn}(LLR_n)}{2}$, where $\text{sgn}(LLR_n)$ is the sign of the LLR_n, we perform the check described by $x \cdot H^T = 0$. If it holds, the decoded word has been derived, and the decoding process is terminated.
- Soft Decision-Core Part
  The difference in this approach lies in the fact that we do not use the entire codeword and the whole matrix H for the check, but just the core part of the matrix. The core part refers to the first 26 columns (variable nodes) and 4 rows (check nodes) for BG1, while for BG2, it pertains to the first 14 columns and 4 rows, outlining the structure relevant to each base graph configuration. The relation is the same $x_c \cdot H_c^T = 0$ but for a specific part x_c of the x and H_c of the H.

2.2 Hardware Requirements Between Core Part and Full H

Table 1 shows the number of check nodes connected to a specific number of variable nodes for the two base graphs, in case we use either the entire H matrix or the core part approach. Specifically, two nodes are considered to communicate if there is 1 at the corresponding position in the parity-check matrix H (columns for variable nodes, rows for check nodes).

The circuit for implementing equation $x \cdot H^T = 0$ in $GF(2)$ consists of parallel XOR gate trees. Each multiplication operation between the vector x and each

Table 2. Needed LUT Based on Method

Method	LUT
Full H	67200
Core Part	17664

Table 3. Packets

TBS	R	Zc	BG
912	120/1024	96	BG2
1480	193/1024	160	BG2
2408	308/1024	256	BG2
4680	602/1024	224	BG1
6144	679/1024	288	BG1

column of the matrix H^T forms a distinct tree. More specifically, after transposing the matrix H, the columns of the matrix correspond to the check nodes, and the rows correspond to the variable nodes. Consequently, the number of GF(2) two-input additions and thus the number of XOR gates needed, is proportional to the number of connections each check node has. Table 1 shows the number of XOR gates required for the implementation of the trees depending on the number of connections, considering the case where $Z_c=1$.

However, the circuit must be capable of handling the worst-case scenario ($Z_c=384$), meaning 384 identical instances trees must be created for each case. Consequently, when using the entire H matrix for the check, where all columns and rows are used, a total of 735×384=282240 XOR gates are required to have a circuit that satisfies both BG1 and BG2. In contrast, when applying the core part logic, only 216×384=82944 XOR gates are needed. As a result, a 70% reduction is achieved. However, the critical path remains the same, as in both cases and for both base graphs, the check node with the highest number of connections (19 for BG1 and 10 for BG2) remains, which sets the critical path to 6 XOR gates for BG1 and 5 XOR gates for BG2. An FPGA has been used for verification, namely a Zynq UltraScale+ MPSoC ZCU104, and the results for the required lookup tables (LUT) are presented in Table 2, revealing a 73% reduction.

2.3 Termination at a Sub-iteration Level

Both of these criteria can be applied at the end of the computation of each layer, i.e., several times during an iteration, rather than only at the end of an iteration. Implementing a simple soft decision check with the full H following each layer evaluation, instead of at the end of an iteration, does not provide a significant advantage, as the number of clock cycles saved typically ranges from 3 to 40 cycles for the architecture in Fig. 4. In contrast, adopting the core part approach at the end of each layer, a notable gain is observed at high Signal-to-Noise Ratio (SNR) levels, approximately around 150 cycles. Additionally, at the iteration level, a moderate gain of about 1.2 iterations is achieved, applicable for the BG1 and BG2 configurations, as illustrated in Figs. 1 and 2 respectively for two packets of Table 3, where TBS is the Transport Block Size and R is the Code Rate. This reduction is achieved by **maintaining the BLER** at exactly the same levels ensuring that the decoding performance remains unchanged.

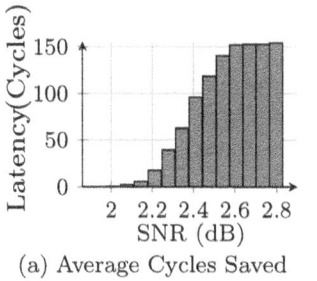
(a) Average Cycles Saved

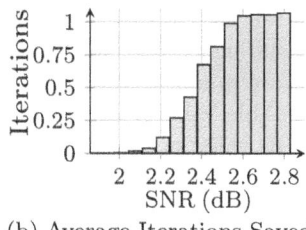
(b) Average Iterations Saved

Fig. 1. Saved Cycles (a) and Iterations (b), TBS=4680, R=602/1024, Z_c=224, BG1

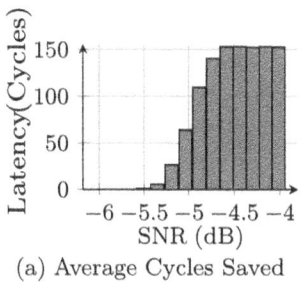
(a) Average Cycles Saved

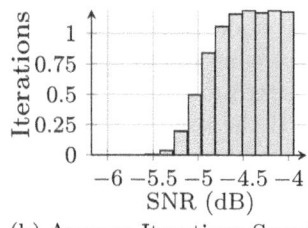
(b) Average Iterations Saved

Fig. 2. Saved Cycles (a) and Iterations (b), TBS=912, R=120/1024, $Z_c = 96$, BG2

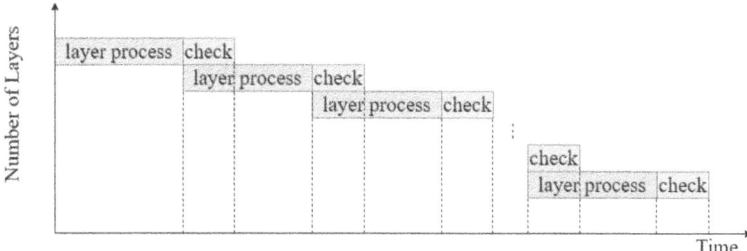

Fig. 3. Pipelined Check and Layer Processing within an Iteration

Since already existing hardware are utilized, no additional complexity is introduced. Furthermore, the maximum delay path of the unit is 6 XOR gates. If it cannot be accommodated at the end of the cycle, it can overlap with the next cycle's processing, adding only a clock cycle for termination, as shown in Fig. 3.

3 Proposed Scheduling Scheme

In this section, we present a different approach to scheduling the checks processed each time, based on identifying the LLR with the minimum value. The objective is to accelerate convergence. The idea is based on the fact that the algorithm uses the minimum value among the received $V2C$ to calculate the $C2V$ message. A lower absolute value of LLR means a greater uncertainty about the correctness of the corresponding estimation. By selecting the variable node with the minimum LLR value, we minimize the impact on the other variable nodes, since only the smallest possible value is subtracted from the estimation, while simultaneously correcting the value of the chosen variable node.

3.1 Approach of Minimum LLR

In this approach, we find the variable node with the smallest non-zero LLR. This is motivated by the fact that during the initial phase of decoding the first two variable nodes are set to zero (puncture bits). Consequently, including zero-valued nodes in the rescheduling logic would not yield significant benefits. Once the variable node with the smallest LLR is identified, we serially process the check nodes connected to it that have not yet been processed during the current iteration. After processing all the particular check nodes, the partial check logic is applied. If the condition is not satisfied, the process continues by selecting the next variable with the smallest LLR excluding the previously processed variable node from further consideration. The process terminates when all the check nodes have been processed within the current iteration.

3.2 Syndrome-Select Logic on Minimum LLR Approach

Building on the previous idea, we propose an additional feature that utilizes the syndrome select logic. In this method, prior to identifying the minimum LLR, we calculate the syndromes by identifying the checks that do not satisfy equation $x \cdot H^T = 0$. Once the variable node with the minimum LLR is determined, we process the checks that are connected with the specific variable node, have not yet been processed, and are not satisfied ($x \cdot H^T \neq 0$). This ensures that check nodes that satisfy the equation $x \cdot H^T = 0$ are excluded from further processing.

3.3 Rescheduling

The approaches described in Sects. 3.1 and 3.2, achieve good decoding performance as shown in Fig. 8; however, the hardware requirements are significant, as the unit to compute the minimum value among the various $LLRs$ it can be very complex. Thus, we propose an algorithm, tested for OMS and NMS, with a different order of processing check nodes, incorporating the logic of partial core check of Sect. 2 and the syndrome-select. The new order of check is presented in Table 4. The order results from the quantization (7 bit signed numbers with 1 bit fractional) and the large possible values of Z_c ($Z_c \leq 384$), which lead to the assumption that the probability of a variable node presenting a minimum LLR is approximately 1, in combination with the existence of puncture bits at the first two variable nodes in the first iteration. Thus, instead of calculating each time the variable node with the smallest absolute LLR value (not zero) and processing the checks that communicate with it, a fixed approach is adopted.

Table 4. Rescheduling

Iteration	BG1	BG2
First	1, 2, 3, 11, 24, 27, 34, 4, 5, 6, 7, 8, 9, 10, 12, 13, 14, 15, 16, 18, 20, 21, 23, 25, 29, 31, 33, 35, 37, 39, 41, 43, 45, 17, 19, 22, 26, 28, 30, 32, 36, 38, 40, 42, 44, 46	4, 1, 23, 25, 27, 29, 31, 34, 37, 41, 3, 10, 16, 20, 38, 5, 6, 8, 9, 11, 13, 14, 15, 17, 18, 21, 32, 36, 39, 42, 2, 7, 12, 19, 22, 24, 26, 28, 30, 33, 35, 40
Second to Tenth	1, 2, 3, 4, 5, 6, 7, 8, 9, 10, 12, 13, 14, 15, 16, 18, 20, 21, 23, 25, 27, 29, 31, 33, 35, 37, 39, 41, 43, 45, 11, 17, 19, 22, 24, 26, 28, 30, 32, 34, 36, 38, 40, 42, 44, 46	1, 2, 3, 5, 6, 7, 9, 11, 12, 14, 16, 19, 20, 22, 24, 26, 28, 30, 33, 35, 37, 40, 4, 8, 10, 13, 15, 17, 18, 21, 23, 25, 29, 32, 36, 39, 42, 27, 31, 34, 41, 38

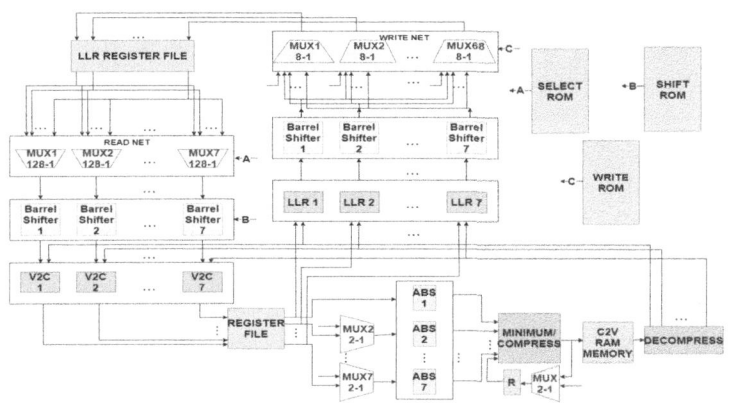

Fig. 4. Semi Parallel Architecture

3.4 Hardware Requirements of Proposed Approach

The hardware required for the introduced rescheduling approach is exactly the same as using the complete H logic for the check. The only difference lies in the order of processing the layers that compose each iteration, which implies a reorganization of ROM memories of architecture in Fig. 4 that controls the select signals within the circuit. This adjustment ensures that the selection signals align with the new processing order of the layers, allowing for correct control flow. The hardware used for the full H logic check is sufficient to support both the core part check, which we adopted in the proposed approach and constitutes part of the overall control mechanism, as well as the syndrome-select logic, which is described by equation $x \cdot H^T = 0$. In terms of circuit-level comparison with the core part approach, the differences are outlined in Sect. 2.2.

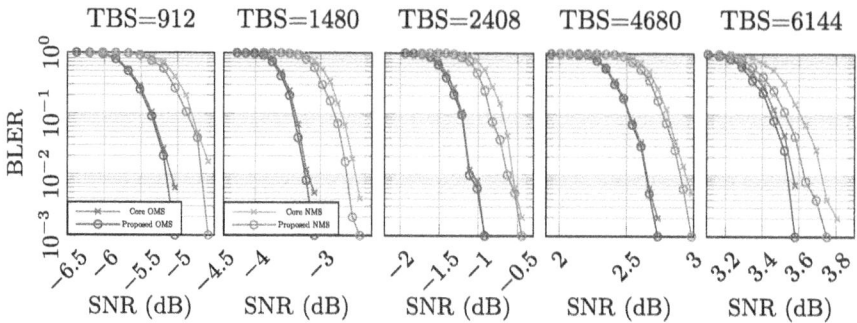

Fig. 5. BLER *vs.* SNR for Different TBS

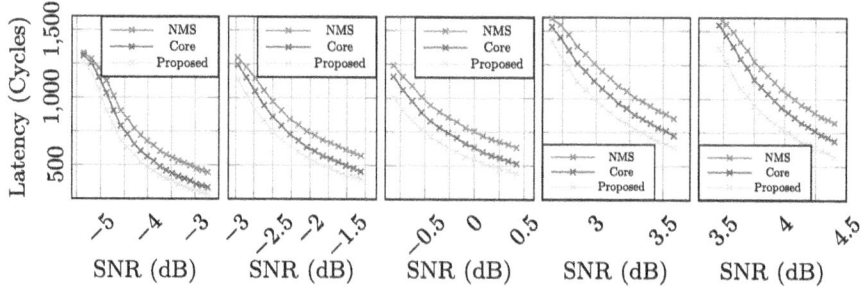

Fig. 6. Latency *vs.* SNR (NMS) for Different TBS

4 Evaluation

This section compares for the OMS and NMS the full-H and the core part check at the end of each layer with the proposed idea. The influence of each approach on the BLER is examined and compared, along with the number of clock cycles for decoding. The algorithm used is OMS with $b=0.5$ and NMS with $a=0.75$, quantization of 7 bits signed numbers with 1-bit fractional part, and a maximum of 10 iterations. The measurements concern the packets of Table 3.

4.1 Block Error Rate

The BLER diagrams comparing the core part logic and the rescheduling approach are presented in Fig. 5. Across all packets, the rescheduling approach slightly outperforms the core part logic, which as mentioned in Sect. 2.3 achieves the same BLER with the full matrix H. Especially for the NMS algorithm the improvement is quite significant. This fact makes the results presented below, regarding the reduction of required cycles, particularly significant.

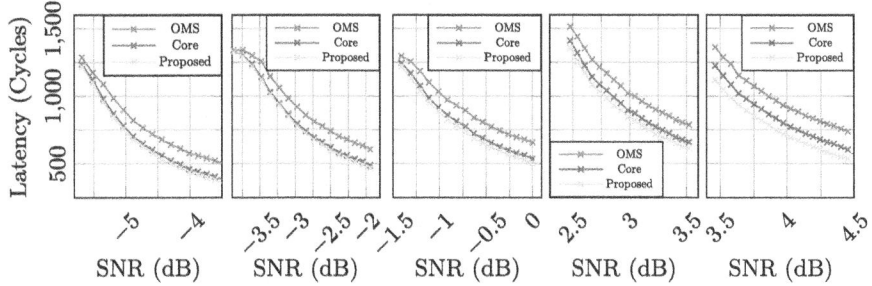

Fig. 7. Latency vs. SNR (OMS) for Different TBS

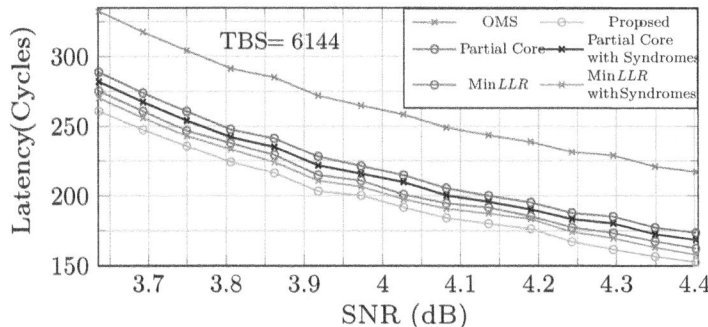

Fig. 8. Latency vs. SNR: Comparison of the Proposed Scheme with Prior Approaches

4.2 Average Needed Cycles for Decoding

Figures 6 and 7 show the average number of clock cycles required for OMS/NMS with full H, the core part, and the proposed approach for the semi-parallel architecture of Fig. 4. It is shown that the new scheduling method reduces the cycles by up to 35%, compared to 18% reduction achieved by the core part logic. The additional reduction in the core part logic, which achieves a reduction of a maximum of 150 cycles, ranges from 20 to 40 cycles for OMS/BG2, 50–110 cycles for OMS/BG1, 45–150 cycles for NMS/BG2 and 100–200 cycles for NMS/BG1. The difference is larger in low SNR where the number of decoding iterations is quite large, reaching a peak of 200–220 cycles when the BLER ranges to 10^{-1}. The average reduction is significant considering that in certain cases the reduction between rescheduling and core part version can be substantial, reaching as high as 560 cycles.

An overview of all the methods studied for a full-parallel architecture for OMS and TBS= 6144 is presented in Fig. 8. While using only the core part of matrix H reduces cycles by 18%, the proposed scheme improves reduction to 25%, 28%, and 35% for the minimum LLR without syndrome logic, with syndrome logic, and rescheduling, respectively. Also the integration of the syndrome-select into the core part approach is presented, resulting in a 20% reduction. This demon-

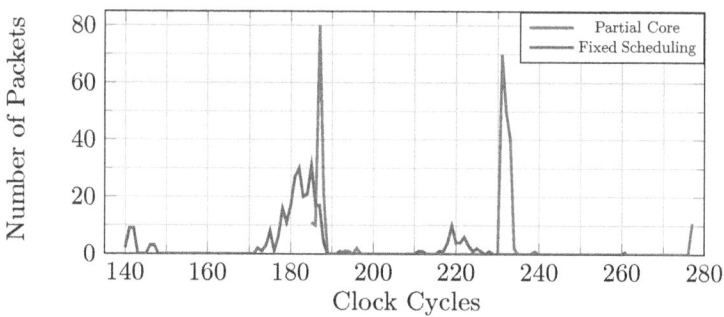

Fig. 9. Packet Dispersion Between Partial Core and Rescheduling (SNR= 4.08 dB)

strates the improvement achieved through the proposed scheme (35% reduction) obtained by combining the rescheduling method with syndrome-select logic.

Furthermore, and for same scenario studied above (i.e., full parallel architecture, TBS=6144, OMS), Fig. 9 illustrates the packet dispersion at a specific SNR=4.08 dB. We observe a gradual shift in the number of packets requiring fewer cycles between the simple partial core logic and the proposed fixed scheduling approach. Specifically, packets that require 230 and 185 cycles in the core part logic are distributed across 190 to 230 cycles and 170 to 180 cycles in the proposed approach, respectively.

4.3 ASIC Implementation - Power Consumption

For the semi-parallel architecture of Fig. 4 which is capable of handling all Z_c and BG, instead of using 19 units at each step [17], i.e., the maximum weight of check nodes, we use 7. This simplification is based on the structure of the BG1/BG2 matrices shown in Table 1, where only 4/0 nodes need three cycles for processing, 5/4 need two, and 37/38 need one cycle. The architecture has been synthesized using an ASIC 28-nm technology and Cadence Genus, excluding the check node RAM, targeting a 10-ns clock period (100 MHz). A power consumption of 0.4 W was reported. As a result, from Sect. 4.2 for a representative case of NMS with TBS=6144 (BG1) and at SNR=4.027 dB, we observe the following average cycles: 1036 for the full H, 901 for the core, and 752 for the proposed approach. Thus, the energy consumption is E_{FH}=0.4 W×1036 cycles×10 ns=4144 nJ for full H, E_{CP}=0.4 W×901 cycles×10 ns=3604 nJ for core part and for rescheduling is E_R=0.4 W×752 cycles×10 ns=3008 nJ. As a result, the required energy is further reduced, from 13% with the core part to 27% with rescheduling.

5 Reconfigurable Barrel Shifter

The n-bit shifter required for the 5G LPDC decoder must support shifts for a range of 54 different Z_c values, making it necessary for the circuit to accommodate shifts on words shorter than n bits. Although the $N \times N$ Benes network [2]

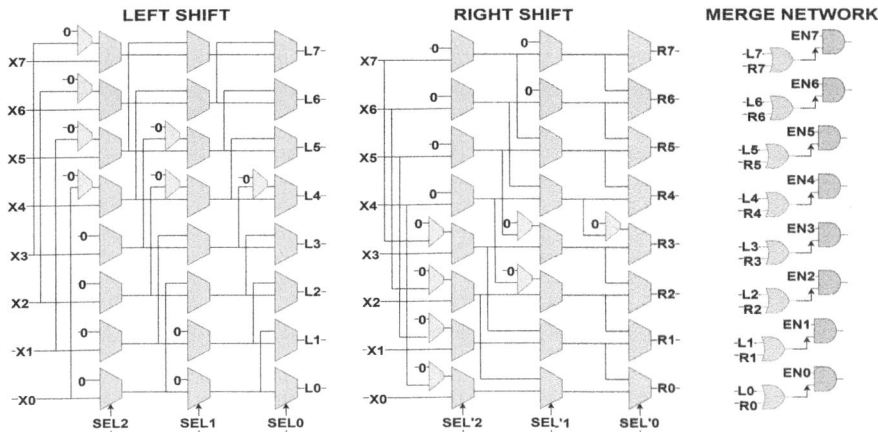

Fig. 10. Reconfigurable Barrel Shifter

can perform any sort of permutations over N input data, where N is a power of two, in our situation we utilize a barrel shifter, as we only need a right rotation.

For the reconfigurable shifter, various ideas have been implemented, one of which is the QSN presented by Chen et al. [5], which serves as a basis for the proposed approach. Specifically, the shift is achieved through a combination of two multiplexer networks for a right and a left shift, with the final result produced by a merge network composed of n multiplexers. On the contrary, self routing network approach is presented by Liu et al. in [15] utilizing only one barrel shifter and consisting of three stages: combination of source messages with the self-routing bits (SRB), barrel shifter and finally the selection scheme. An improved version is presented in [13] capable to route the messages associated with different sub-matrices concurrently. A variation of design in [15], is presented by Liu et al. in [14], where the concept of SRB is not utilized. Instead, it incorporates the addition of two parallel routing decision circuits after the barrel shifter, while maintaining a selection scheme. A different logic, tailored for cases with a smaller number of configurations, is presented by Tsatsaragkos and Paliouras in [20] where only right rotate takes place introducing additional multiplexers at intermediate stages to enable shifts for various word lengths.

Here, we propose an alternative approach to implement the merge network, leveraging the fact that positions without data default to a zero value. Consequently, after the right and left shift from two barrel shifter networks of 384 bits, instead of a selection circuit utilizing 384 multiplexers, the output can be produced with a group of 384 OR gates between the two shifted signals. When it is necessary to maintain zero values in empty positions, as in our LDPC decoder, an additional mask comprising an AND gate can be added to ensure the retention of zeroes in these positions, as illustrated in Fig. 10. Additionally, the circuit can handle multi-frames by placing multiplexers at appropriate points, as shown in Fig. 10. In this specific case, the circuit can handle two frames with a size less than or equal to 4 bits. To derive the enable signals of the mask, a look-up table is necessary. Since for each of the 54 possible Z_c values, 384 enable signals are required, an initial approach would suggest a memory size of 54 × 384=20736 bits. However, as consecutive positions across Z_c values share the same selection signals, the required memory size is reduced to 54×54=2916 bits.

6 Conclusion

A core part check at a sub-iteration level is proposed, which leads to a significant reduction in the number of layers processed, reaching up to 150 clock cycles, maintaining the BLER at the exactly same level, while simultaneously reducing the hardware area. Moreover, by altering the scheduling of the checks processed each time, a new scheduling approach is presented, as mentioned in Sect. 3.3, which leads to a reduction in clock cycles, equal to 35%, and on power consumption. Simultaneously the BLER is improved without any additional hardware. Finally, a different merge network on shifter has been shown to simplify the multiplexers and reduce the critical path.

Acknowledgement. This work has been supported by the Horizon Europe Project X-TREME 6G: X Transceivers & RF front-ends made in Europe's Microelectronics light house to Enable new 6G use cases (X-TREME 6G - 101192681).

References

1. 3rd Generation Partnership Project: 3GPP - The Mobile Broadband Standard (2023). https://www.3gpp.org/about-3gpp. Accessed 01 Jun 2024
2. Beneš, V.E.: Optimal rearrangeable multistage connecting networks. Bell Syst. Tech. J. **43**(4), 1641–1656 (1964). https://doi.org/10.1002/j.1538-7305.1964.tb04103.x
3. Casado, A.I.V., Griot, M., Wesel, R.D.: LDPC decoders with informed dynamic scheduling. IEEE Trans. Commun. **58**(12), 3470–3479 (2010). https://doi.org/10.1109/TCOMM.2010.101910.070303
4. Chen, J., Fossorier, M.: Density evolution for two improved BP-Based decoding algorithms of LDPC codes. IEEE Commun. Lett. **6**(5), 208–210 (2002). https://doi.org/10.1109/4234.1001666
5. Chen, X., Lin, S., Akella, V.: QSN-A simple circular-shift network for reconfigurable quasi-cyclic LDPC decoders. IEEE Trans. Circuits Syst. II Express Briefs **57**(10), 782–786 (2010). https://doi.org/10.1109/TCSII.2010.2067811
6. Emran, A.A., Elsabrouty, M.: Simplified variable-scaled min sum LDPC decoder for irregular LDPC codes. In: 2014 IEEE 11th Consumer Communications and Networking Conference (CCNC), pp. 518–523 (2014).https://doi.org/10.1109/CCNC.2014.6940497
7. Fossorier, M., Mihaljevic, M., Imai, H.: Reduced complexity iterative decoding of low-density parity check codes based on belief propagation. IEEE Trans. Commun. **47**(5), 673–680 (1999). https://doi.org/10.1109/26.768759

8. Gallager, R.: Low-density parity-check codes. IRE Trans. Inf. Theory **8**(1), 21–28 (1962). https://doi.org/10.1109/TIT.1962.1057683
9. Hocevar, D.: A reduced complexity decoder architecture via layered decoding of LDPC codes. In: IEEE Workshop on Signal Processing Systems, 2004. SIPS 2004, pp. 107–112 (2004). https://doi.org/10.1109/SIPS.2004.1363033
10. Kim, J.H., Nam, M.Y., Song, H.Y., Lee, K.M.: Variable-to-check residual belief propagation for informed dynamic scheduling of LDPC codes. In: 2008 International Symposium on Information Theory and Its Applications, pp. 1–4 (2008). https://doi.org/10.1109/ISITA.2008.4895535
11. Kim, N.I., Kim, J.U.: Early termination scheme for 5G NR LDPC Code. In: 2021 International Conference on Information and Communication Technology Convergence (ICTC), pp. 933–935 (2021). https://doi.org/10.1109/ICTC52510.2021.9621117
12. Li, J., He, G., Hou, H., Zhang, Z., Ma, J.: Memory efficient layered decoder design with early termination for LDPC codes. In: 2011 IEEE International Symposium of Circuits and Systems (ISCAS), pp. 2697–2700 (2011). https://doi.org/10.1109/ISCAS.2011.5938161
13. Liu, C.H., Lin, C.C., Chang, H.C., Lee, C.Y., Hsu, Y.: Multi-mode message passing switch networks applied for QC-LDPC decoder. In: 2008 IEEE International Symposium on Circuits and Systems (ISCAS), pp. 752–755 (2008). https://doi.org/10.1109/ISCAS.2008.4541527
14. Liu, C.H., et al.: Design of a multimode QC-LDPC decoder based on shift-routing network. IEEE Trans. Circuits Syst. II Express Briefs **56**(9), 734–738 (2009). https://doi.org/10.1109/TCSII.2009.2027967
15. Liu, C.H., et al.: An LDPC decoder chip based on self-routing network for IEEE 802.16e applications. IEEE J. Solid-State Circ. **43**(3), 684–694 (2008). https://doi.org/10.1109/JSSC.2007.916610
16. MacKay, D.: Good error-correcting codes based on very sparse matrices. IEEE Trans. Inf. Theory **45**(2), 399–431 (1999). https://doi.org/10.1109/18.748992
17. Nguyen Ly, T.T.: Efficient hardware implementations of LDPC decoders, through exploiting impreciseness in message-passing decoding algorithms. Theses, Université de Cergy Pontoise (2017). https://theses.hal.science/tel-01783859
18. Petrović, V.L., El Mezeni, D.M.: Reduced-complexity offset min-sum based layered decoding for 5G LDPC codes. In: 2020 28th Telecommunications Forum (TELFOR), pp. 1–4 (2020). https://doi.org/10.1109/TELFOR51502.2020.9306590
19. Shao, R., Lin, S., Fossorier, M.: Two simple stopping criteria for turbo decoding. IEEE Trans. Commun. **47**(8), 1117–1120 (1999). https://doi.org/10.1109/26.780444
20. Tsatsaragkos, I., Paliouras, V.: A reconfigurable LDPC decoder optimized for 802.11n/ac applications. IEEE Trans. Very Large Scale Integr. (VLSI) Syst. **26**(1), 182–195 (2018). https://doi.org/10.1109/TVLSI.2017.2752086
21. Xu, M., Wu, J., Zhang, M.: A modified Offset Min-Sum decoding algorithm for LDPC codes. In: 2010 3rd International Conference on Computer Science and Information Technology. vol. 3, pp. 19–22 (2010). https://doi.org/10.1109/ICCSIT.2010.5564884

Comparison Between In-Core Hardware IDS, Off-Core Hardware IDS and Software IDS

Tianxu Li[✉] ⓘ, Mohamed El-Bouazzati ⓘ, Camille Monière ⓘ, Philippe Tanguy ⓘ, and Guy Gogniat ⓘ

Université Bretagne-Sud, UMR 6285, Lab-STICC, 56100 Lorient, France
tianxu.li@univ-ubs.fr

Abstract. Wireless attacks targeting the Internet of Things (IoT) pose challenges to its security. To counter this threat, in-depth security mechanisms such as Intrusion Detection Systems (IDSs) are used. The implementation of IDSs in edge devices is challenging, considering the inherent constrained nature of IoT devices. In this paper, three Intrusion Detection System (IDS) implementation approaches, software, in-core hardware, and off-core hardware are defined and compared, using an IoT-context representative case study. Advantages and disadvantages of each approach are assessed and discussed, comparing design time, ease of maintenance, detection performance, and SoC resource consumption. Our results, relative to the SoC baseline, show that the software approach used 17.92% more energy consumption per packet ($+0.19 mJ/p$) than the hardware approach. Conversely, the hardware approach incurs a higher FPGA resource overhead, requiring up to 12.06% more LUT and 7.75% more FF.

Keywords: Embedded System Security · Intrusion Detection System · Internet of Things · Wireless Security

1 Introduction

The Internet of Things (IoT) refers to a network of connected physical devices gathering and sharing data to ease everyday tasks and increase productivity. However, as the number of devices continuously grows, security matters become more pressing, especially in wireless communication. Threats like jamming, replay attacks, and memory corruption threaten communication confidentiality, device security and data integrity. Therefore, protecting IoT devices from these threats is crucial to ensure the reliability of the IoT [16].

Detecting threats or attacks is a task commonly assured by an Intrusion Detection System (IDS). It is a security mechanism designed to monitor networks or systems for malicious activities. It analyses data using techniques like signature-matching and anomaly-based detection, gathering information at various levels of the IoT device: hardware, software and network. When a potential threat is detected, the IDS sends an alert to network administrators, or to

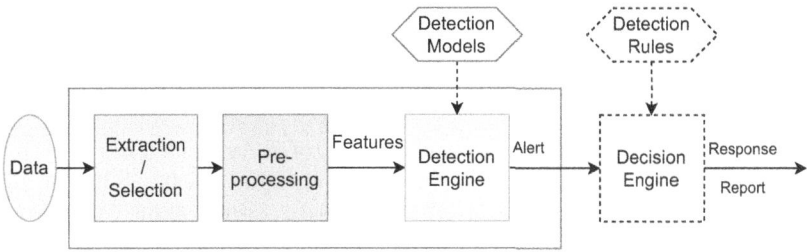

Fig. 1. Functional Model of an IoT-targeting IDS

automated systems (Security Operations Center), enabling a rapid response to mitigate or even prevent the attacks [18].

This paper aims to compare several implementation variations of an on-edge IDS targeting resource-constrained IoT devices, which can detect various types of wireless attacks on low-rate wireless communication protocols. We evaluate the IDS performances in the context of the LoRa physical layer, using a system composed of COTS devices and a RISC-V soft-core processor. The IDS can be implemented in software or in hardware. Each variant has its strengths and limitations, suitable for different network environments and security requirements. To better understand how these implementations perform, this paper quantitatively compares several key performance indicators such as detection performance, SoC resource consumption and energy-efficiency. This comparison will help in evaluating the efficiency and practicality of the different implementations in real-world situations, thereby providing a scientific basis for designing more efficient embedded IDSs and thus, more reliable and secure IoT systems.

In Sect. 2, the context and current studies on IDS architecture and implementation are reviewed. In Sect. 3, the threat model, the IoT-targeting IDS chosen as a case study are summarized, and the three implementation approaches explored are detailed. Section 4 presents the experimental setup and the results, and discusses the design choices based on key performance indicators. Finally, Sect. 5 provides conclusions and future perspectives.

2 Background and Related Works

Securing network communications, particularly through IDSs, is addressed in the literature [9] indeed. Figure 1 describes the flowchart and the main components of an IDS.

This work focuses on the design choices for implementing three key functions: *Extraction/Selection* (E/S), *Preprocessing* (PPC) and the *Detection Engine* (DE). Indeed, most studies focus on optimizing detection models and algorithms implemented in the detection engine or selecting the best features to enhance performance [7,12]. In addition to this, resource-constrained IoT devices often lack the processing power for efficient IDS implementation, necessitating hybrid strategies that offload computations to the cloud [13,15]. This approach can degrade throughput and network performance, especially in systems that prioritize flexibility to support various IoT protocols, as well as widening the attack surface. IDS implementations can be broadly separated into two categories, software and hardware.

Software implemented IDSs process the monitored data and make decisions in software executed by programmable circuits, leveraging the flexibility of microprocessors. They are widely adopted due to their configurability, short implementation time, and lack of need for specialized hardware [3]. However, they can put pressure on the CPU and consume more memory resources than hardware implementations, especially with heavy data traffic or complex attacks, which is challenging for resource-constrained IoT nodes. Nevertheless, Cayre et al. [1] proposed OASIS, a framework to implement an IDS for Bluetooth Low Energy (BLE) devices. The firmware of the BLE controller is instrumented to get features and detection modules are implemented to detect five low-level attacks. It is also extensible to support other attacks. However, the detection modules are based on heuristics, with each module targeting a specific attack. Therefore, it requires more resources to detect additional attacks.

In this paper, the term "hardware implemented IDSs" is used to describe two distinct types of design: (a) a standalone unit that integrates all the functionalities of an IDS, and (b) a unit that performs some tasks of an IDS. Considering this definition, a hardware unit commonly comes in two forms described thereafter, depending on where it is implemented in a System-on-chip (SoC). One is implemented outside the CPU core as an independent peripheral (therefore referred to as "Off-core"), connected to the SoC via Direct Memory Access (DMA) or through the on-chip interconnect bus. It can operate either autonomously or as an accelerator assisting the processor. For instance, hardware accelerators for neural networks assist processors in calculations [11], while IDSs such as [14] enforce security policies independently. Another approach consists in integrating IDSs directly into the processor [5] (therefore referred to as "In-core"). Since they process and analyse data directly at the hardware level, hardware IDSs offer faster response time, lower latency, and minimal utilization of the IoT system computing resources. Thus, existing hardware implementations are designed to allow high-speed network connections without compromising energy efficiency. However, hardware IDSs have higher implementation costs, longer development time, and are harder to modify or upgrade compared to software solutions. Zareen et al. [20] proposed an in-core solution for botnet detection that uses static and unique feature selection and extraction. They evaluated it for both classification accuracy and hardware implementation. El-Bouazzati et al. [2] proposed an IDS for LoRa devices that can detect ongoing remote attacks in real time, and that can be embedded in resource-constrained devices. It uses Hardware Performance Counters (HPCs) to collect data at the processor level and can detect memory corruption attacks. The feature extraction and selection module supports only one data source, i.e., the HPCs, but they can count different events. The implementation has been deployed on an FPGA and evaluated for detection performances and resource usage on a RISC-

V-based SoC. However, the implementation supports only one attack type and lacks extensive evaluation in execution time and energy consumption.

There is currently a lack of research on the comparison of the different IDS implementation approaches, especially for detecting wireless attacks in constrained devices. To the best of our knowledge, no studies specifically discuss hardware, software and hybrid implementations of IDS in this context. Therefore, a quantitative analysis is highly needed to highlight challenges like energy consumption, computation, and memory usage. Additionally, the ability to update the IDS to counter new attacks is an important metric. Regarding IDS selection, we revisited a previously mentioned work [2]. This IDS employs a hybrid signature- and anomaly-based algorithm using a decision tree, a method also utilized in other IDSs. Therefore, this research holds significant representativeness, while being open-source and not complex to re-use. For a fair comparison, we reimplemented and extended the in-core IDS from [2] and applied the same algorithm using software and off-core methods. Our main evaluation criteria include resource consumption, response time, flexibility, energy consumption, and adaptability, with a particular focus on comparing in-core and off-core approaches.

3 Case Study Description

In this section, we introduce our case study by defining the threat model. Then, we succinctly present the selected IDS, its analysing and detection methods, and how it responds to the threat model. Finally, we detail the three considered implementation approaches.

3.1 Threat Model

In this study, the attacker is assumed capable of conducting wireless attacks on embedded devices participating in a LoRa Wide Area Network (LoRaWAN®). The attacker can use dedicated COTS devices to launch attacks remotely, but also have access to versatile Software-defined Radio (SdR) devices. We assume the attacker cannot have physical access to the target device. However, the attacker can manipulate any layer of a communication stack, from the physical layer to the application layer.

Hessel et al. [6] provided an extensive review of LoRaWAN® vulnerabilities and the attacks exploiting them. They found that jamming attacks are common, either disrupting wireless signals directly or enabling more complex attacks like device impersonation. Such attacks can lead to man-in-the-middle scenarios, data leakage, or even complete network takeover. In their paper, the authors categorize jamming attacks based on the attacker's network knowledge and ability to gather nodes information. "Triggered jamming", where the jammer activates only when a preamble is detected, is particularly complex and stealthy. Due to the long transmission time of LoRaWAN® packets, attackers have ample opportunity to detect specific preambles and emit jamming signals during a single transmission. This type of attack is the focus of our study.

Given the increasing number of vulnerabilities in IoT protocol implementations [17], we also consider memory corruption attacks [4] for completeness. These attacks target the implementation of network protocol stacks. They consist in exploiting a memory weakness or vulnerability (such as a buffer overflow on a memory location) to erase a function's return address. This can destabilize programs, leading to denial of service or even to a Remote Code Execution (RCE). In our case, we consider two types of memory corruption attacks, those involving stack overflow and those involving heap overflow.

The IDS takling those threats is described thereafter.

3.2 IDS Architecture and Detection Methods

The IDS considered in this work integrates the one developed in [2], which can be assimilated to a Machine-Learning (ML)-based hybrid signature-based and anomaly-based IDS, leveraging HPCs as data source. It has been proven capable of detecting memory corruption attacks, and distinguishing between stack and heap overflow. In the generic model in Fig. 1, the data would be the events generated by the core. Event selection is done offline, and extraction is performed by storing and retrieving data from HPCs. The decision engine is a decision tree. This approach may be less efficient with a context-switching software platform, like a real-time OS, but is sufficient for bare-metal software running on limited-resource devices.

A HPC is intended to enable developers to monitor processor events, offering insights into the system's performances during runtime. However, several studies have taken advantage of HPCs capabilities to develop robust security mechanisms, particularly for intrusion detection. The processor used in this study includes a clock cycle counter, a "instruction retired" counter, and 29 configurable counters capable of monitoring various events. Previous studies tested ten metrics under memory corruption attacks using a machine learning classifier and identified two key metrics: BRANCH-TAKEN, which counts branch instructions, and LD-STALL, which counts delayed load instructions. A decision tree model has been built for these two metrics, for detection and classification purposes. In our re-implementation, we built a new dataset and retrained the model to get decision tree parameters that fit our research environment.

The IDS can be reduced to two main components: the metric collection module and the decision-making module. The metric collection module configures the HPCs to monitor the two selected events, activating them during the processing of LoRaWAN® data packets and stopping them afterward. The decision-making module receives data from the HPCs and uses a decision tree to detect anomalies. In our upcoming comparative studies, these two modules will be deployed in various ways depending on the IDS implementation, described in Subsect. 3.3.

To improve the representativeness of the IDS, a jamming detector based on statistical analysis of Received Signal Strength Indicator (RSSI) of received LoRaWAN® packets has been added. This extension allows detecting jamming attacks [19]. In the generic model in Fig. 1, the data are the LoRa packets. Extraction consists in retrieving the RSSI from the COTS device, and the value is preprocessed through an Exponentially-Weighted Moving Average (EWMA) filter. The decision engine checks that the resulting value is between upper and lower thresholds. The EWMA filter allows reducing spurious values impact, following the formula: $y_n = \frac{1}{4}x_n - \frac{3}{4}y_{n-1}$, with y_n the EWMA value for the nth received packet, and x_n the RSSI value of the nth received packet. The thresholds are obtained by statistical analysis of the values for legitimate packets over a trustworthy channel. Similar to the IDS designed for memory corruption attacks, this extended IDS for jamming attacks also consists of three key modules: RSSI collection, EWMA preprocessing, and a threshold-based decision module. As previously, these modules will be implemented differently based on EWMA implementation variants, described in Subsect. 3.3.

The IDS described in this section, thereafter simply referred to as "the IDS", should have low power and low compute resource requirements, while still addressing the challenges of the considered threat model. The next section is dedicated to the description of the three implementation variants of the IDS.

3.3 Considered Implementation Approaches

In this paper, we propose to study three kinds of implementations, (1) a software implementation $\mathcal{I}_{SW}$, (2) an in-core hardware one $\mathcal{I}_{ICH}$, and (3) an off-core hardware implementation $\mathcal{I}_{OCH}$. This section describes how the focused blocks in Fig. 1 are dispatched between the software, the CPU circuit, and eventual peripherals as shown in Fig. 2. Data are not represented to lighten the figure, since all approaches use the same, i.e., RSSI values reported by LoRa COTS devices retrieved in software, and HPC values retrieved directly from the CPU.

In the software approach $\mathcal{I}_{SW}$, the extraction (S/E), the preprocessing (PPC), and the decision engine (DE) are run within a program executed by the CPU. In our case, this is achieved using a bare-metal program written in C language. As implied in Fig. 2, it is the simplest approach from a design point-of-view, since all components live in the same domain. However, latency and throughput should be strongly impacted since any access to the hardware is constrained by API calls. Moreover, this approach is subject to common software weaknesses, whether they come from coding malpractice, or result from CPU vulnerabilities [10].

The $\mathcal{I}_{ICH}$ variant can be considered the conceptual opposite. All the IDS functions are implemented as hardware components of the CPU itself, only S/E is partially present in software, due to the requirement of the data source. This requires intimate knowledge of the CPU ISA, and to have access to its hardware description. However, this approach should have a low resource-access overhead, since all components can be tightly coupled, as well as the highest performance and energy efficiency.

In the same fashion, the $\mathcal{I}_{OCH}$ variant is also completely implemented in hardware, except for the software required by the data source. It takes the form of a peripheral connected to the CPU using available SoC interconnections. While still requiring hardware development, it is more independent of the core. It still requires some components to be available in the core, like HPCs in our case, but

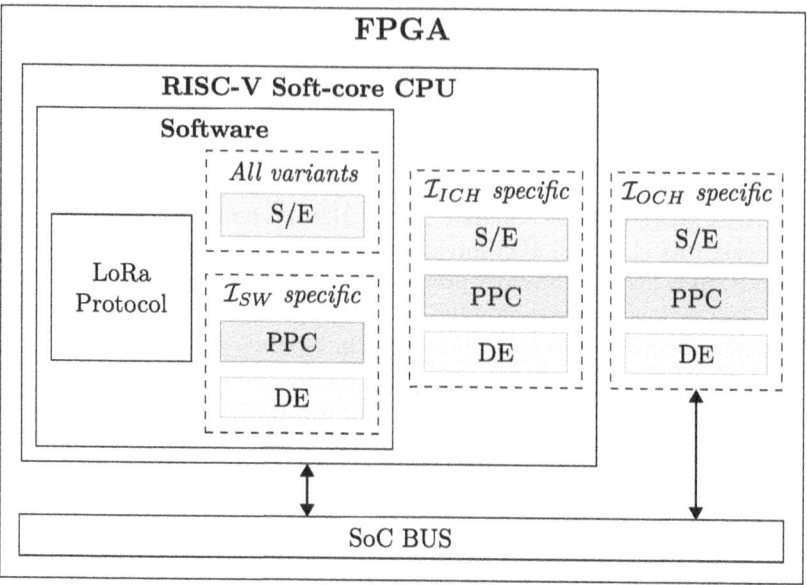

Fig. 2. Selection/extraction (S/E), preprocessing (PPC), and decision engine (DE) function implementation sites (software, CPU logic and peripheral logic) depending on the implementation variant.

it is a common requirement of mainstream ISAs like RISC-V, ARM, or x86_64. It should be less efficient than $\mathcal{I}_{ICH}$ due to extra latency introduced by the bus, but still more than $\mathcal{I}_{SW}$ as it is not overly reliant on API calls.

To quantify their respective strength, a test platform and protocol has been defined and is described in the next section.

4 Results and Discussion

This section outlines the experimental setup, then it presents the results that enabled us to compare each approach. Before concluding, we provide insights into the advantages and disadvantages of each method.

4.1 Test Bench

To fairly compare the three IDS implementations, we ensured identical experimental conditions for each implementation, using the protocol shown in Fig. 3. Test data are replayed by the SdR through shielded cables to an IDS-enabled device, with adjusted gain to ensure consistent Signal-to-Noise Ratio (SNR). The IDS-enabled device is composed of a LoRa shield COTS device connected to an Artix 7 Xilinx FPGA (xc7a100t) using SPI. In the FPGA, a RISC-V RV32IMC processor (CV32E41P) is deployed. RISC-V allows modifying freely the core description. The hardware is described in System Verilog, and the complete SoC is generated using LiteX [8], an open-source Python library that streamlines hardware integration.

Exclusively used for evaluation, the test data consist of I/Q samples captured by mimicking full-scale scenarios, insuring that the IDS's behaviour closely mirrors actual deployment conditions. For each attack type considered, we recorded 400 packets using a SdR device and a LoRa Shield at a distance of 33 meters with direct line-of-sight. This included 200 genuine packets and 200 packets attacked by an independent LoRa COTS device.

This test bed allows us to verify the detection performances for each IDS implementation. Since we used the same IDS model across all three architectures, performance indicators like precision, accuracy, and F1 score are consistent. The packet loss rate is approximately 2% for legitimate packets and memory corruption attack packets, and about 50% for jamming attack packets, thus resulting in 26% in average. Only successfully decoded packets were accounted, since the IDS cannot detect packets that are not received, as stated in Sect. 3.2. Nevertheless, for jamming attacks, the performance metrics of the three IDS architectures are equivalent. For the memory corruption attack, the result also showed an identical detection performance, with all attacks successfully detected and no false positives. This outcome aligns with our expectations: IDSs using the same model yield the same (or nearly identical) results across different architectures.

4.2 Results

This section compares the resource consumption of different IDS architectures to provide quantitative reference results for selecting the appropriate deployment strategy.

First, we focus on the hardware resource consumption required by the IDS, reported in Table 1, and obtained after place-and-route and bitstream generation using the synthesis tool (namely, Xilinx Vivado 2022). For the software variant $\mathcal{I}_{SW}$, the only additional hardware resources consumed compared to the basic SoC correspond to the overhead induced by adding HPCs to the RISC-V core, a common requirement of all IDS implementation approaches. Unsurprisingly, hardware IDS variants consume significantly more hardware resources than the software one. The $\mathcal{I}_{OCH}$ consumes slightly more hardware resources than the $\mathcal{I}_{ICH}$ because it requires extra logic links on the interconnect bus to transmit monitored data. However, this increase is minimal - only 0.41% more FF and

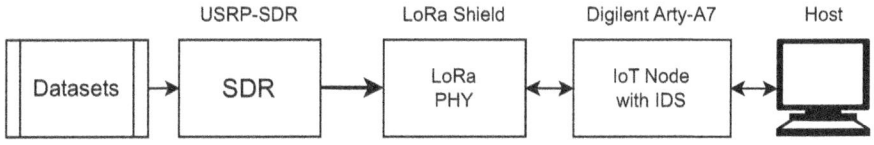

Fig. 3. Dataset replay setup to test the IDS

Table 1. Resource consumption and maximum achievable CPU frequency (CMF) without IDS ($\mathcal{I}_{base}$) with the software IDS ($\mathcal{I}_{SW}$), the in-core IDS ($\mathcal{I}_{ICH}$), or the off-core IDS ($\mathcal{I}_{OCH}$) on an Artix 7 FPGA (xc7a100t).

Implem.	Soft-core CPU's FPGA resources					SoC (*Freq:* 50 *MHz*)			
	LUTs		FFs		CMF	LUTs		FFs	
	Value	Cost (%)	Value	Cost (%)	MHz	Value	Cost (%)	Value	Cost (%)
$\mathcal{I}_{base}$	4676	N/A	2136	N/A	65.69	4800	N/A	7887	N/A
$\mathcal{I}_{SW}$	4777	2.16	2217	3.79	65.60	4883	1.72	8044	1.99
$\mathcal{I}_{ICH}$	5345	14.30	2625	22.89	65.07	5283	10.06	8466	7.34
$\mathcal{I}_{OCH}$	4777	2.16	2217	3.79	65.60	5379	12.06	8498	7.75

2.00% more LUT compared to the $\mathcal{I}_{ICH}$. This is because even though the $\mathcal{I}_{ICH}$ is integrated within the CPU core, it still requires additional connections to transmit monitored indicators and receive results, consuming resources similar to those used by a lightweight interconnect bus. The difference in hardware resources is insignificant for the entire SoC. However, modifying the core to add new IDS modules and new connections supposes access to the hardware description, rarely provided with closed-source ISA like ARM.

We also consider the maximum operating frequency of the IDS. For $\mathcal{I}_{SW}$, its maximum frequency theoretically matches that of the CPU (CPU Maximum Frequency, CMF). From the perspective of the overall system's maximum operating frequency, there isn't much difference. This is because the hardware IDS we selected and built is lightweight and high-speed, with the $\mathcal{I}_{ICH}$ operating maximum at 218.05 MHz and the $\mathcal{I}_{OCH}$ maximum at 220.60 MHz. The SoC's timing constraints primarily stem from other components, like the core and peripherals.

The usage of software resources will also vary depending on the IDS architecture. Even for hardware IDS, the in-core and off-core architectures will have different resource consumption due to the varying software support needed for collecting metrics, transmitting metrics, and obtaining results. Since we used bare-metal programming without a Real-Time Operating System (RTOS) in the resource-constrained IoT nodes, we measured software resource consumption by the size of the compiled binary files. As shown in Table 2, the code size increases significantly with the addition of the $\mathcal{I}_{SW}$, resulting in a 7.64% increase for the resource-constrained IoT node. However, the specific increase of 1566 bytes is acceptable for our lightweight IDS, but may be more impactful with complex IDS implementations. $\mathcal{I}_{ICH}$ has the smallest increase in code size, only 0.58%, while $\mathcal{I}_{OCH}$ requires an additional link to interconnect bus, slightly increasing the code size by 342 bytes. However, it still offers advantages in memory space compared to $\mathcal{I}_{SW}$.

Next, we analysed the required clock cycles, including the IDS detection time (the number of cycles from receiving metrics to deciding) and the complete time (the total cycles from packet reception to processing completion). The number of clock cycles for all complete times and $\mathcal{I}_{SW}$ processing times were obtained using

Table 2. IDS implementation effect on firmware size and CPU clock cycles

IDS Implem.	Size (bytes)	Cost		Processing Cycles		
		Direct (bytes)	Relative (%)	Jamming Detection	Mem. Corr. Detection	Complete
$\mathcal{I}_{base}$	20,500	∅	∅	∅	∅	112,028
$\mathcal{I}_{SW}$	22,066	1,566	7.64	1,034	227	133,215
$\mathcal{I}_{ICH}$	20,618	118	0.58	3	1	112,457
$\mathcal{I}_{OCH}$	20,842	342	1.67	4	1	113,099

the core's built-in cycle counter. This counter starts when a packet is received or detection begins and stops at the end of processing. To ensure stability, we began measurements after processing 10 packets, then recorded the complete times for the next 30 packets and calculated the average. For the processing times of the two hardware IDS architectures, we used Vivado simulation software, building a test platform in System Verilog to obtain exact processing cycles through simulation and waveform observation.

In terms of the clock cycles required for data processing, both hardware IDS architectures are much faster than the software IDS. Regarding the complete cycles, the total processing time does not significantly increase for hardware IDS because they can handle metrics for jamming attacks and memory corruption attacks in parallel. On the other hand, the software IDS needs to execute sequentially, and with the added overhead of CPU control, the number of cycles of complete time increases significantly. $\mathcal{I}_{ICH}$ has an advantage over $\mathcal{I}_{OCH}$ in terms of metric transmission time, but this does not significantly impact the total time.

In practice, IoT nodes will not reach saturation due to the duty cycle regulations. Even the slower software IDS can complete processing before the next packet arrives. However, it is important to note that, for some time-sensitive applications, prolonged processing times may lead to response delays, causing the response packets to either be missed or rejected.

4.3 Discussions

Considering the previous results, several conclusions can be drawn. Both $\mathcal{I}_{ICH}$ and $\mathcal{I}_{OCH}$ variants require, on average, around +7.54% (+595) of FFs and +11.06% (+531) of LUTs compared to the SoC baseline. Activating the HPCs adds an extra overhead of +1.99% (+157) of FFs and +1.72% (+83) of LUTs, consistent across all IDS versions. Despite this, hardware IDS implementation eliminates the need for additional memory usage in an IoT device. In contrast, the software IDS requires an additional +7.63% (1566 bytes) of memory compared to the original firmware. It's also important to note that hardware solutions generally have longer design times than software ones.

The CPU Maximum Frequency (CMF) remains stable at around 65.60 MHz across all versions, unaffected by the implementation approach. The in-core and off-core can reach higher maximum frequencies (up to 218.05 MHz and 220.60 MHz) when operating independently. While the SoC reached 50.00 MHz due to other timing constraints, the use of another clock domain would allow benefiting from the in-core and off-core IDS maximum frequency. However, the software IDS operates at the standard SoC frequency, which means it will be slower, especially when handling more complex IDS tasks.

The processing time and, therefore, the response time, varies between approaches. The software IDS requires more time (+1261 clock cycles in total) to detect memory corruption attacks and jamming attacks, compared to the others, which require no more than 5 clock cycles to detect these attacks. In Table 2, the last column includes the number of clock cycles required to completely process a packet, which also accounts for the control required by each approach. We observe that this time is significantly higher in the software IDS, which is +18.91% (+21,187 clock cycles) relative to the baseline processing time, corresponding to 423.74 μs of response time. The response time reaches 21.42 μs for the software version, offering more opportunities for a successful attack compared to hardware solutions.

Finally, energy consumption estimates using the Xilinx Vivado power estimation tool show an average power consumption of 470 mW for the SoC in all approaches. As a comparison point, we calculated the coarse-grained energy consumption per packet (mJ/p) by multiplying the processing cycles of each approach by the package power consumption and dividing by the SoC, clock frequency (50 MHz). The baseline energy consumption per packet is 1.06 mJ/p for both hardware implementations. However, it reaches 1.25 mJ/p for the software IDS due to additional CPU usage and processing cycles. Since the hardware resources consumed by our hardware-based IDS are negligible compared to the overall complexity of the SoC, which includes the processor and other peripherals, the primary factor influencing energy consumption is the number of clock cycles utilized. Thus, the results highlight the higher energy efficiency of hardware approaches compared to the software one, which can directly impact the overall battery lifetime of an IoT device.

5 Conclusion and Perspectives

Through a detailed comparison of different implementation approaches for an IoT IDS, in terms of hardware and software resource consumption, processing time, detection performance, and estimation of energy efficiency, we have identified distinct advantages and disadvantages for each approach. Hardware approach (both in-core and off-core) indeed demonstrates significantly faster processing speeds, thus reducing energy consumption and response time. Despite their higher resource consumption, requiring more LUTs and F, for lightweight IDSs, the benefits outweigh the costs. The in-core IDS has the lowest overall resource requirements, while the off-core IDS requires slightly more resources but is way more portable, as it is not entangled with the CPU microarchitecture. Software IDS, however, excels in flexibility and portability, allowing easy updates and configurations to counter new threats.

Among hardware options, the off-core approach thus stands out compared to in-core IDS due to its unique advantages. This dual capability to maintain efficient performance while offering flexibility and scalability makes it the suitable choice for the integration and expansion of current systems. Future work will focus on the study of more complex IDS for advanced cores such as CVA6 with embedded operating systems such as Zephyr RTOS. Portable IP-based IDS libraries with partial reconfiguration support are also targeted, to help developers improve security measures efficiently.

References

1. Cayre, R., Nicomette, V., Auriol, G., Kaâniche, M., Francillon, A.: OASIS: an intrusion detection system embedded in Bluetooth low energy controllers. In: Proceedings of the 19th ACM Asia Conference on Computer and Communications Security (ASIA CCS). ACM. https://doi.org/10.1145/3634737.3645004
2. El Bouazzati, M., Tessier, R., Tanguy, P., Gogniat, G.: A lightweight intrusion detection system against IoT memory corruption attacks. In: Proceedings of the IEEE International Symposium on Design and Diagnostics of Electronic Circuits and Systems (DDECS). IEEE. https://doi.org/10.1109/DDECS57882.2023.10139718
3. Eskandari, M., Janjua, Z.H., Vecchio, M., Antonelli, F.: Passban IDS: an intelligent anomaly-based intrusion detection system for IoT edge devices. https://doi.org/10.1109/JIOT.2020.2970501
4. Github, Inc.: NVD - CVE-2022-39274. https://nvd.nist.gov/vuln/detail/CVE-2022-39274
5. Harris, A., et al.: Morpheus II: A RISC-V security extension for protecting vulnerable software and hardware. In: Proceedings of the IEEE International Symposium on Hardware Oriented Security and Trust (HOST). https://doi.org/10.1109/HOST49136.2021.9702275
6. Hessel, F., Almon, L., Hollick, M.: LoRaWAN Security: An Evolvable Survey on Vulnerabilities, Attacks and their Systematic Mitigation. https://doi.org/10.1145/3561973
7. Jan, S.U., Ahmed, S., Shakhov, V., Koo, I.: Toward a Lightweight Intrusion Detection System for the Internet of Things. https://doi.org/10.1109/ACCESS.2019.2907965
8. Kermarrec, F., Bourdeauducq, S., Lann, J.C.L., Badier, H.: LiteX: An open-source SoC builder and library based on Migen Python DSL. https://doi.org/10.48550/arXiv.2005.02506
9. Khraisat, A., Alazab, A.: A critical review of intrusion detection systems in the Internet of Things: techniques, deployment strategy, validation strategy, attacks, public datasets and challenges. Cybersecurity **4**(1), 1–27 (2021). https://doi.org/10.1186/s42400-021-00077-7
10. Kocher, P., et al.: Spectre attacks: Exploiting speculative execution. https://doi.org/10.1145/3399742

11. Ngo, D.M., Temko, A., Murphy, C.C., Popovici, E.: FPGA hardware acceleration framework for anomaly-based intrusion detection system in IoT. In: Proceedings of the 31st International Conference on Field-Programmable Logic and Applications (FPL). https://doi.org/10.1109/FPL53798.2021.00020
12. Oh, D., Kim, D., Ro, W.W.: A Malicious Pattern Detection Engine for Embedded Security Systems in the Internet of Things. https://doi.org/10.3390/s141224188
13. Pongle, P., Chavan, G.: Real Time Intrusion and Wormhole Attack Detection in Internet of Things. https://doi.org/10.5120/21565-4589
14. Pontarelli, S., Bianchi, G., Teofili, S.: Traffic-Aware Design of a High-Speed FPGA Network Intrusion Detection System. https://doi.org/10.1109/TC.2012.105
15. Raza, S., Wallgren, L., Voigt, T.: SVELTE: Real-time intrusion detection in the Internet of Things. https://doi.org/10.1016/j.adhoc.2013.04.014
16. Schiller, E., Aidoo, A., Fuhrer, J., Stahl, J., Ziörjen, M., Stiller, B.: Landscape of IoT security. https://doi.org/10.1016/j.cosrev.2022.100467
17. Siwakoti, Y.R., Bhurtel, M., Rawat, D.B., Oest, A., Johnson, R.C.: Advances in IoT Security: Vulnerabilities, Enabled Criminal Services, Attacks, and Countermeasures. https://doi.org/10.1109/JIOT.2023.3252594
18. Soniya, S.S., Vigila, S.M.C.: Intrusion detection system: Classification and techniques. In: Proceedings of the International Conference on Circuit, Power and Computing Technologies (ICCPCT). https://doi.org/10.1109/ICCPCT.2016.7530231
19. Zahra, F.T., Bostanci, Y.S., Soyturk, M.: Real-Time Jamming Detection in Wireless IoT Networks. https://doi.org/10.1109/ACCESS.2023.3293404
20. Zareen, F., Fernandes Amador, M.A., Karam, R.: Hardware Immune System for Embedded IoT. https://doi.org/10.1109/TCSII.2022.3187312

Comparative Study of Memory Optimization Techniques for Dataflow-Modeled Applications

Naouel Haggui[1], Maxime Pelcat[1], Yaesop Lee[2], Shuvra Bhattacharyya[1(✉)], Kevin Martin[3], and Jean-François Nezan[1]

[1] University of Rennes, INSA Rennes, CNRS, IETR - UMR 6164, 35000 Rennes, France
{nhaggui,mpelcat,jnezan}@insa-rennes.fr, ssb@umd.edu
[2] Kwangwoon University, Seoul, South Korea
yaesop@terpmail.umd.edu
[3] Université Bretagne-Sud, CNRS UMR 6285, Lab-STICC, Lorient, France
kevin.martin@univ-ubs.fr

Abstract. Efficient memory management is essential for signal and image processing systems, particularly in data-intensive applications where performance and resource constraints are critical. This paper presents a comparative study of two advanced memory optimization techniques: Memory Script Optimization (MSO), and Passive Active Flow Graph (PAFG) Optimization-within the context of dataflow-modeled applications. Both approaches aim to reduce memory usage and improve execution efficiency, but they do so with distinct strategies: Memory Scripts focus on in-place buffer management, while PAFG modifies actor interactions to minimize buffer requirements. Using a portion of a Convolutional neural network (CNN) application as a case study, we evaluate the efficiency of these techniques in terms of memory reduction and execution time. Our results demonstrate that MSO provides significant performance improvements, achieving up to 17% memory savings and 21% faster execution times, making it ideal for independent data operations. However, PAFG offers greater scalability and flexibility, particularly when dealing with complex data dependencies, and provides a simpler path to implementation. This work not only highlights the trade-offs between memory efficiency and flexibility but also paves the way for applying these optimizations in near-memory computing architectures, where distance from memory to processing is employed as a parameter to improve efficiency.

Keywords: PAFG · Memory script · Dataflow

1 Introduction

The rapid expansion of data-intensive applications, particularly in image and signal processing with the introduction of Machine Learning (ML) in most workloads, has underscored the critical importance of memory optimization in embedded systems. As these applications grow in complexity and scale, the demand for efficient memory management techniques becomes increasingly crucial. Multiprocessor Systems-on-Chips (MPSoCs) are widely employed to execute such applications, yet they often encounter significant limitations in terms of internal memory capacity and bandwidth. Optimizing memory usage not only enhances performance but also determines the feasibility of deploying these applications in resource-constrained environments like embedded systems and edge computing devices.

Synchronous Dataflow (SDF) modeling is frequently employed in signal processing systems due to its ability to represent applications as directed graphs, naturally expressing parallelism and ensuring deterministic execution [2]. In these graphs, computational units called actors exchange data through first-in, first-out (FIFO) buffers. While this model efficiently describes data dependencies and execution flows, traditional dataflow approaches often struggle with managing memory effectively, particularly when handling inter-actor communication and large-scale data storage, leading to significant performance bottlenecks [3].

To tackle these challenges, this paper investigates two advanced memory optimization techniques from state-of-the-art: Memory Script Optimization (MSO) [3] and Passive Active Flow Graph (PAFG) Optimization [6]. Both methods aim to minimize memory usage and improve execution efficiency by employing distinct strategies for buffer management. MSO focuses on merging input and output buffers using script-based definitions, drawing on a detailed understanding of internal data dependencies and manual sub-buffer reuse strategies to enable in-place computation [3]. On the other hand, PAFGs transform active actors into passive forms by replacing traditional actor firing with read/write operations to a locally shared address space, simplifying the dataflow structure and reducing memory overhead [6].

In this paper, we compare these two optimization techniques using a portion of the SqueezeNet application. SqueezeNet is a CNN-based image classification application designed to achieve the accuracy of AlexNet while utilizing significantly fewer parameters. This makes it adopted for resource-constrained environments like embedded systems. We provide an analysis of the PAFG and MSO impact on both memory usage and execution speed, highlighting the trade-offs between the two approaches. The study aims at providing insights for developers to choose the most suitable optimization method based on the specific requirements of their application, balancing memory efficiency, parallelism, and ease of implementation. Moreover, this work takes a step towards applying these memory optimization techniques in near-memory computing architectures [10]. Near-memory computing consists of moving computation closer to memory, reducing the latency and energy costs associated with transferring data between memory and processing units. Traditional computing architectures suffer from the *memory wall*, where data transfer between memory and processing cores becomes a bottleneck, especially for data-heavy applications. Near-memory computing addresses this issue by integrating processing elements nearby memory, reducing data movement overhead and improving performance for memory-bound applications [9,10]. By applying MSO and PAFG in such architectures, we can further streamline memory usage and execution efficiency, making dataflow-modeled applications more suitable for real-time, performance-constrained environments.

The rest of this paper is organized as follows: Sect. 2 reviews existing memory optimization techniques, while Sect. 3 provides a detailed explanation of the logic behind memory script and PAFG optimizations. Section 4 presents a comparative analysis of both approaches, and Sect. 5 concludes the paper.

2 Background

Dataflow modeling is used in image and signal processing due to its support for determinism, parallelism, and the high-level structural representation of applications [2]. A key challenge in the development of data-intensive applications is managing memory efficiently, as it directly impacts performance. Optimizing memory usage has been the focus of various research efforts, particularly in applications modeled using dataflow graphs. In [1,11], FIFO sizing techniques were introduced to minimize the memory footprint in SDF applications by finding a schedule to minimize the memory space allocated for each FIFO of an SDF graph. In [4], Geilen et al. propose a method for optimizing token storage by considering token production and consumption as synchronized events. This proposal makes it possible to reuse the same memory space for tokens that are produced and consumed during the execution of an actor. In [7], other techniques such as annotation systems have been presented to enable buffer merging but are limited to pairwise merging and monocore architectures.

In 2016, a new optimization technique that outperforms the state-of-the-art techniques, called MSO, has been introduced [3]. It aims to minimize the memory footprint by reusing buffers and aligning data allocation with internal data dependencies of the application. This method allows for significant memory reduction while maintaining performance, as it optimizes buffer allocation strategies across multiple actors in the dataflow model. Another approach that has been introduced recently in the dataflow field is the PAFG method [6]. PAFG restructures the dataflow to minimize buffer requirements and eliminate unnecessary data duplication. It also improves execution performance by streamlining data transfer between actors. In [5], Ghasemi et al. investigated efficient memory management strategies to overcome cache limitations in dataflow applications. They proposed two cache-optimized techniques-Copy-on-Write (CoW) and Non-Temporal Memory Copying (NMC)-which demonstrated notable performance gains, particularly in multicore environments. Their experiments showed that these approaches could significantly reduce cache miss rates for the Stereo application, with reductions of up to 60% in the best case, and could improve application speed by up to 2.10×, all without requiring changes to the actor's source code. While PAFG and MSO focus on optimizing the dataflow structure at compile time to reduce memory usage and improve execution efficiency, the techniques presented by Ghasemi et al. [5] are complementary, targeting runtime optimization through cache management strategies. These approaches aim to improve cache miss rates and data transfer efficiency, enhancing performance at runtime without altering the dataflow model itself.

This paper focuses on a comparison of the PAFG and MSO approaches. Additional details on each approach, including their memory-saving techniques and performance impact, are provided in Sect. 3.

3 MSO and PAFG

This section details the compared methods.

3.1 Memory Script Optimization

The MSO presented in [3], is a technique to minimize the memory footprint of DSP applications modeled with SDF graphs on shared-memory MPSoCs. This technique exploits a manual description of the actor, internal data dependencies (actor port annotations) and script-based specifications for merging input and output buffers, achieving up to a 48% memory footprint reduction on a set of examples compared to state-of-the-art methods. The proposed method explores memory reuse possibilities by managing contiguous memory spaces called buffers, referenced by pointers. This approach avoids complex pointer operations and enhances code readability. Input and output buffers, though logically separate, are often mapped to the same memory space to enable memory reuse. New information on actor behavior and buffer access is required to enable memory reuse. This method has been implemented in a dataflow tool called PREESM [8]. PREESM offers a variety of actors. Some of those actors such as the Fork, Join, Broadcast, and Roundbuffer are called special actors as they manage data tokens without performing computations. Those special actors can automatically benefit from MSO because scripts are provided within the tool. For example, Broadcast actors copy input buffer content to multiple output buffers, allowing memory allocation to be reduced by merging buffers. User-defined actors can also benefit from advanced memory reuse techniques by revealing disjoint lifetimes of buffer sub-ranges. Corrupting an input buffer by writing results into it is acceptable if no live data is overwritten, where live data refers to tokens that have been produced but have not yet been consumed by the receiving actor.

Figure 1 illustrates an example of a memory script for the Broadcast actor. The memory script for the broadcast actor enables efficient memory usage by mapping each output buffer directly to the input buffer without duplicating data. This is achieved by applying the `matchWith` function, which aligns each output buffer (`bOut[i]`) with the entire range of the input buffer (`bIn`). As a result, all outputs access the same memory location, reducing memory allocation requirements. This approach minimizes runtime overhead and preserves memory resources by eliminating redundant data copies across outputs.

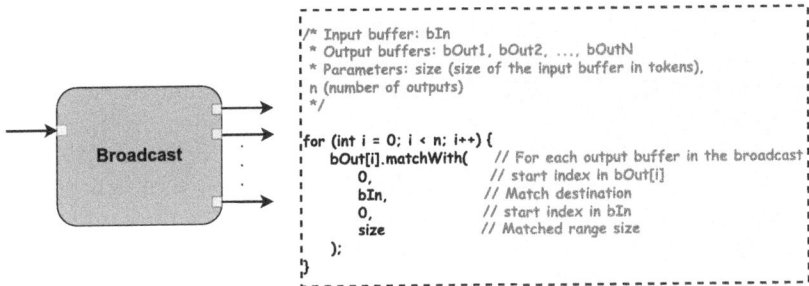

Fig. 1. Memory script for the Broadcast actor

3.2 PAFGs Optimization

In signal processing systems, PAFGs offer advantages by optimizing memory usage, particularly by reducing buffer requirements and preventing unnecessary data duplication. This makes them suitable for processing large datasets and complex signal-processing tasks. PAFGs enhance execution efficiency by minimizing the computational overhead associated with actor firing and data transfer, leading to faster execution times and improved system performance-critical factors in real-time applications [6]. PAFGs work by converting active actors into passive ones. In traditional dataflow models, an active actor performs computations through enable/invoke interfaces. In contrast, passive actors come in two forms: simple and non-simple. A simple passive actor operates similarly to a standard buffer in a dataflow system, while the non-simple passive actor mirrors an active actor by mapping input streams to output streams but operates via read/write interfaces instead of the typical enable/invoke approach.

Figure 2 illustrates a transformation of the Broadcast actor from an active to a passive form. In the original active form shown in Fig. 2(a), the Broadcast actor has one input and two outputs. During each execution, the actor consumes a single token from its input and generates a single token for each output, creating identical copies of the input token on each output. Figure 2(b) provides pseudocode for this active Broadcast actor, highlighting the overhead associated with duplicating the input value for both outputs. This duplication introduces a runtime cost, due to the time required to copy data between the input and output FIFOs, as well as an increased memory requirement. This overhead is necessary under a strict dataflow interpretation, where the input must be replicated across both output edges (FIFOs). In contrast, Fig. 2(c) depicts the passive form of the Broadcast actor, which diverges from the pure dataflow model. Rather than operating through a series of discrete firings, the passive Broadcast functions similarly to a typical FIFO component. Here, a buffer is associated with the component, and tokens are managed using a write pointer and two read pointers - one for each output edge of the Broadcast actor. This design eliminates the need to fire the actor actively; instead, tokens are written to and read from the buffer ports as needed. By adding a second read pointer, each output can independently access the input token without requiring the active copying process. This transformation shifts the Broadcast actor from an active firing model to a passive component, thereby reducing runtime overhead and memory demand.

The efficiency of the PAFG approach has been evaluated on three distinct applications: Jitter Measurement, Channelizer, and Error Vector Magnitude computation running on an Intel Core i7-2.5 GHz system. Results demonstrate that PAFG can significantly reduce redundant token storage, improve throughput by approximately 16%, and lower buffer memory requirements by around 25%, highlighting its potential for enhancing performance in real-world signal processing tasks [6].

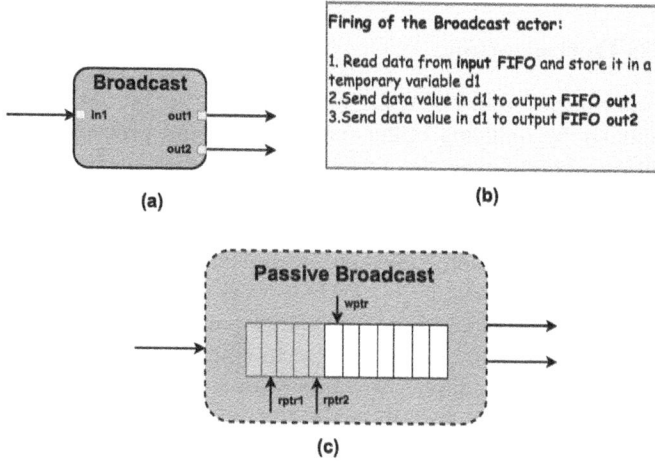

Fig. 2. (a)Broadcast actor. (b) Pseudocode fragment for the Broadcast actor. (c) Passive version of Broadcast actor; wptr, rptr1, and rptr2 show possible positions of the write pointer and the two read pointers.

4 Comparing MSO and PAFG on SqueezeNet Maxpool Optimization

To compare the two approaches outlined in Sect. 3, a portion of the SqueezeNet application (see Fig. 3) is used. Both approaches are applied to the maxpool actor within SqueezeNet. As illustrated in Fig. 4, the SqueezeNet process begins with the loadImage actor, which supplies an input image of size $W0 \times H0$ to the convolution actor. This actor performs a convolution on the image, producing 64 feature maps (or convolution layers), each with dimensions $W1 \times H1$. Maxpooling is then applied to each feature map independently, exploiting the parallelism of the maxpool calculations since the convolution layers are independent. The maxpool actor reduces the sizing of the input data by selecting the maximum value within a defined window.

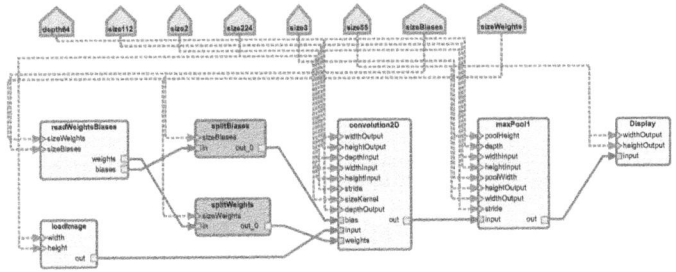

Fig. 3. A subset of SqueezeNet dataflow graph

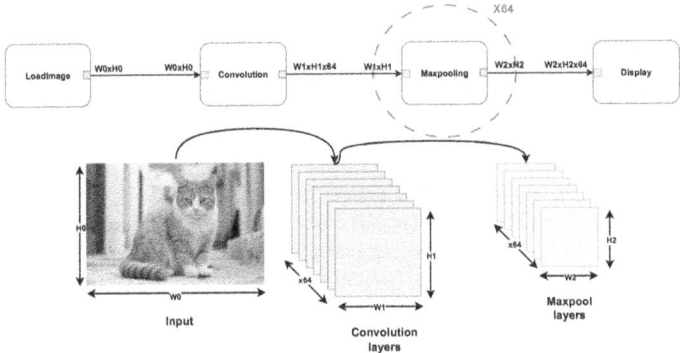

Fig. 4. SqueezeNet: Maxpooling layers parallelism

To expose this parallelism in the dataflow, the developer employs production and consumption rates. Figure 4 demonstrates this by showing that the convolution actor has a production rate of $W1 \times H1 \times 64$, which are then consumed by the maxpooling actor at a rate of $W1 \times H1$ at a time. This structure leads to the execution of the maxpooling actor 64 times, along with 64 corresponding FIFO buffers between the convolution and maxpool actors, and another 64 FIFO buffers between the maxpooling and display actors. Although the dataflow graph of SqueezeNet appears to have only seven FIFOs, when exploiting full graph parallelism, there are a total of 133 FIFOs due to actor duplication. To implement MSO, a memory script is developed specifically for the maxpool actor to optimize memory allocation and management during the maxpooling process. This memory script ensures efficient use of memory for storing pooled feature maps, reducing the computational load while retaining the key features of the input data. By contrast, to apply the second approach, the PAFG method, the maxpool actor is transformed into a passive form. This is achieved by replacing the actor with read and write functions.

Figures 6 and 5 present the algorithms for the read and write functions.

Algorithm 1: maxPooling_write: Applies max pooling and writes results to input buffer

---

Input: Global parameters (widthInput, heightInput, etc.) and input data pointer
Output: Max pooled data stored back in the input buffer
if $input = null$ **then**
  Print "ERROR: Null pointer passed as input.";
  return
if $widthInput \leq 0$ **or** $heightInput \leq 0$ **then**
  Print "ERROR: Invalid size for input layer.";
  return
if $poolWidth \leq 0$ **or** $poolHeight \leq 0$ **then**
  Print "ERROR: Invalid size for max pool operator.";
  return
Allocate $tempOutput$ of size $widthOutput \times heightOutput$;
for $x = 0$ **to** $widthInput - (poolWidth - 1)$ **step** $stride$ **do**
  for $y = 0$ **to** $heightInput - (poolHeight - 1)$ **step** $stride$ **do**
    $max \leftarrow$ Perform max pooling computation on window;
    $tempOutput[(x/stride) + (y/stride) * widthOutput] \leftarrow max$;
for $i = 0$ **to** $widthOutput \times heightOutput - 1$ **do**
  $input[i] \leftarrow tempOutput[i]$;
Free $tempOutput$;

Fig. 5. The `maxPooling_write` algorithm performs max pooling on input data and stores the results back in the input buffer.

The `maxPooling_read` function updates the input pointer for the max pooling operation, while the `maxPooling_write` function performs the max pooling computation by iterating over the input data and writing the results directly back into the FIFO buffer.

From the user's perspective, the dataflow graph remains unchanged (Fig. 4), as the maxpooling task is still represented in the graph. However, from the developer's perspective, the maxpooling actor is no longer an active actor but has been replaced with an empty block, essentially part of the FIFO itself. A new read/write interface is created to substitute the original maxpooling function. This interface enables the computation to be performed within the FIFO, effectively reducing the number of FIFOs from 128 to 64.

Figure 7 illustrates this transformation. In Fig. 7 (a), the duplication of the maxpool actor is shown, where each convolutional output connects to a unique FIFO before reaching the maxpool actor, and another FIFO is placed between the maxpool and display actors. This configuration results in a total of 128 FIFOs. However, with PAFG, as shown in Fig. 7 (b), the read and write functions operate directly on the FIFOs, allowing only 64 FIFOs to achieve the same

Algorithm 2: maxPooling_read: Sets the input pointer for max pooling operation

Input: Input data pointer *in*
Output: Global input pointer updated for max pooling
Update input pointer: ;
$input \leftarrow in$;

Fig. 6. The maxPooling_read algorithm

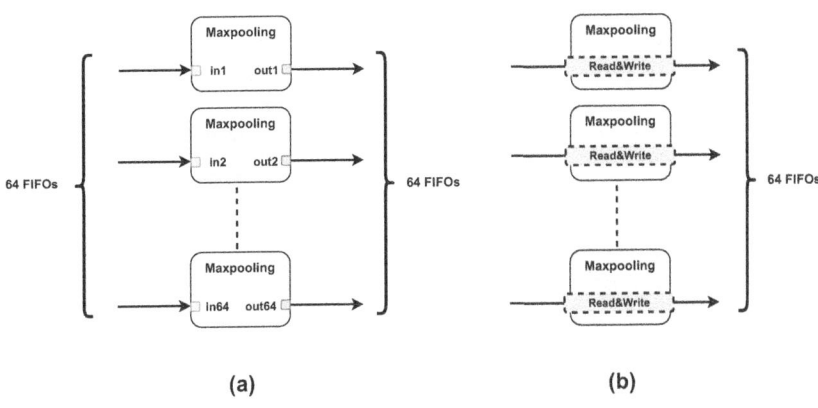

Fig. 7. Maxpool and FIFOs Duplication: (a) Without PAFG. (b) With PAFG

functionality by processing data within the FIFO buffer itself. This approach mirrors in-place computation, similar to MSO.

Table 1 presents the results of applying both optimizations to the maxpool actor across 1-, 2-, and 4-core architectures. The maxpool memoryscript approach demonstrates the most significant improvements, achieving approximately 17% memory reduction and 16–21% faster execution times compared to the reference. On the other hand, the passive maxpooling approach offers more moderate gains, with about 5–6% memory reduction and 6–7% faster execution times.

Although the MSO has proven to be more effective than the PAFG method on the present example, it presents limitations in some situations. Applying MSO sometimes forces developers to choose between using optimization and maintaining parallelism. In the studied case, parallelism was introduced between different feature maps, which are independently computed. However, in scenarios like the one shown in Fig. 8, where the maxpooling function is applied to different slices of the same feature map, there are significant dependencies between the slices. The green blocks in Fig. 8 represent the areas shared between different slices. For example, the numbers 2 and 6 are shared between the first and second slices. While MSO can be applied in such cases, the developer faces the challenge of finding the best way to schedule the execution of the different threads. For instance, in the case presented in Fig. 8, where an in-place operation is com-

Table 1. Comparison of Memory and Time Consumption for Different Optimization Methods Across Multiple Cores

Cores	Method	Time (ms)	Total memory usage (KB)	Memory Reduction (%)
1 Core	Reference	18	17 124	–
	PAFG	17	16 128	5.8%
	MSO	15	14 212	16.9%
2 Cores	Reference	15	17 224	–
	PAFG	14	16 324	5.2%
	MSO	12	14 300	16.9%
4 Cores	Reference	14	17 288	–
	PAFG	13	16 416	5.0%
	MSO	11	14 400	16.7%

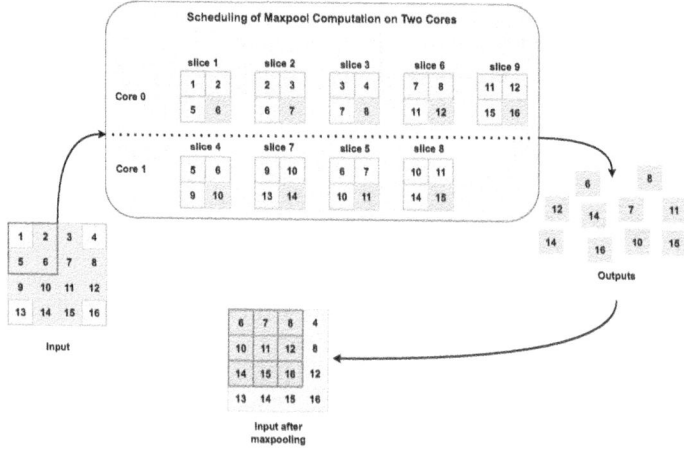

Fig. 8. Maxpooling slice parallelism: In-place computation

puted using the memory script on a 2-core architecture and each slice (1 to 9) corresponds to a thread, the developer must carefully consider the scheduling. Threads 1 and 4 are executed first, as they are independent, followed by threads 2 and 7, which depend on threads 1 and 4, respectively. Then, threads 3 and 5 are executed, relying on threads 2, 1, and 4. Afterward, threads 6 and 8 are scheduled, depending on threads 5 and 7, with thread 9 being executed last, depending on thread 8.

In a nutshell, the results demonstrate that while MSO can deliver the most significant memory and execution time improvements, the PAFG method offers a more flexible solution, as it enables developers to embed computations within passive blocks, set up shared memory spaces, and cluster actors within a common address space-features that are crucial when managing dependencies across data slices, such as those in Fig. 8. PAFG allows overlapping of computation and memory management, letting slices that share data process simultaneously without excessive data duplication or strict scheduling constraints. The choice between these two methods depends on the specific application requirements, with PAFG offering a flexible solution when exploiting parallelism in the target platform is an important objective. Table 2 summarizes the advantages and disadvantages of both MSO and PAFG, providing a clear comparison of their performance characteristics and suitability for different application requirements.

Table 2. Advantages and Disadvantages of PAFG and MSO

Method	Advantages	Disadvantages
MSO	- Provides significant memory savings (up to 17%) and faster execution times (16–21%) - Supports efficient in-place computations, reducing memory allocation overhead - Works well for applications with independent data operations	- Requires detailed scripting and actor annotations, increasing implementation complexity - Can be challenging to maintain parallelism in scenarios with inter-dependent data - Limited flexibility when dealing with complex data dependencies
PAFG	- Offers flexibility by transforming active actors into passive ones, simplifying dataflow - Reduces buffer requirements and unnecessary data duplication - Enhances parallelism and memory sharing, suitable for complex data dependencies	- Achieves more modest memory savings (5–6%) and execution speedup (6–7%) compared to MSO - Requires changes in actor functionality, such as using read/write interfaces instead of active firing

5 Conclusion

This paper evaluates two memory optimization techniques, Memory Script Optimization (MSO) and Passive Active Flow Graph (PAFG) Optimization, on a SqueezeNet maxpool actor optimization. The Memory Script approach delivers better overall performance, achieving up to 17% memory savings and 16–21% faster execution times, especially in tasks involving independent data operations like maxpooling across feature maps. It excels in enabling efficient in-place computations, making it highly effective for streamlining memory usage in data-intensive environments. In contrast, PAFG Optimization offers greater flexibility by converting active actors into passive ones, simplifying the dataflow structure, and reducing buffer requirements. While its memory and execution time improvements are more modest (5–6% and 6–7%, respectively), it proves valuable for handling complex data dependencies, such as when processing slices of the same feature map.

References

1. Benazouz, M., Marchetti, O., Munier-Kordon, A., Urard, P.: A new approach for minimizing buffer capacities with throughput constraint for embedded system design. In: ACS/IEEE International Conference on Computer Systems and Applications-AICCSA 2010, pp. 1–8. IEEE (2010)
2. Bhattacharyya, S.S., Deprettere, E.F., Leupers, R., Takala, J.: Handbook of signal processing systems. Springer (2013)

3. Desnos, K., Pelcat, M., Nezan, J.F., Aridhi, S.: On memory reuse between inputs and outputs of dataflow actors. ACM Trans. Embed. Comput. Syst. (TECS) **15**(2), 1–25 (2016)
4. Geilen, M., Basten, T., Stuijk, S.: Minimising buffer requirements of synchronous dataflow graphs with model checking. In: Proceedings of the 42nd Annual Design Automation Conference, pp. 819–824 (2005)
5. Ghasemi, A., Cataldo, R., Diguet, J.P., Martin, K.J.: On cache limits for dataflow applications and related efficient memory management strategies. In: Workshop on Design and Architectures for Signal and Image Processing (14th edition), pp. 68–76 (2021)
6. Lee, Y., Liu, Y., Desnos, K., Barford, L., Bhattacharyya, S.S.: Passive-active flowgraphs for efficient modeling and design of signal processing systems. J. Sig. Process. Syst. **92**, 1133–1151 (2020)
7. Murthy, P.K., Bhattacharyya, S.S.: Buffer merging-a powerful technique for reducing memory requirements of synchronous dataflow specifications. ACM Trans. Des. Autom. Electron. Syst. (TODAES) **9**(2), 212–237 (2004)
8. Pelcat, M., Desnos, K., Heulot, J., Guy, C., Nezan, J.F., Aridhi, S.: Preesm: a dataflow-based rapid prototyping framework for simplifying multicore DSP programming. In: 2014 6th European Embedded Design in Education and Research Conference (EDERC), pp. 36–40. IEEE (2014)
9. Singh, G., et al.: A review of near-memory computing architectures: opportunities and challenges. In: 2018 21st Euromicro Conference on Digital System Design (DSD), pp. 608–617. IEEE (2018)
10. Singh, G., et al.: Near-memory computing: past, present, and future. Microprocess. Microsyst. **71**, 102868 (2019)
11. Stuijk, S., Geilen, M., Basten, T.: Exploring trade-offs in buffer requirements and throughput constraints for synchronous dataflow graphs. In: Proceedings of the 43rd Annual Design Automation Conference, pp. 899–904 (2006)

Author Index

A
Archet, Agathe 43
Asiyabi, Reza Mohammadi 69

B
Bhattacharyya, Shuvra 121
Bianchi, Edoardo 81
Busia, Paola 57

D
Demo, Nicola 81

E
Ebrahimiazandaryani, Farhad 3
El-Bouazzati, Mohamed 108

F
Fey, Dietmar 3

G
Gac, Nicolas 43
Gogniat, Guy 108
Gorgoń, Marek 28

H
Haggui, Naouel 121

K
Krichene, Hana 16
Kryjak, Tomasz 28

L
Lee, Yaesop 121

Lezoray, Olivier 69
Li, Tianxu 108
Lis, Konrad 28

M
Martin, Kevin 121
Meloni, Paolo 57
Meneghetti, Laura 81
Monière, Camille 108

N
Nazir, Saqib 69
Nezan, Jean-François 121
Nguyen, Le Nam Hieu 16

O
Orieux, François 43

P
Paliouras, Vassilis 95
Papageorgiou, Nikos 95
Pelcat, Maxime 121
Pinna, Andrea 57

R
Rozza, Gianluigi 81

T
Tanguy, Philippe 108

V
Ventroux, Nicolas 43

The manufacturer's authorised representative in the EU is Springer Nature Customer Service Centre GmbH, Europaplatz 3, 69115 Heidelberg, Germany. If you have any concerns regarding our products, please contact ProductSafety@springernature.com

Printed and bound by CPI Group (UK) Ltd, Croydon, CR0 4YY

25/03/2026

02078191-0019